# Intra-Site Obsidian Distribution and Consumption Patterns in Northern Belize and the North-Eastern Peten

Helen R. Haines
Michael D. Glascock

BAR International Series 2326
2012

Published in 2016 by
BAR Publishing, Oxford

BAR International Series 2326

*Intra-Site Obsidian Distribution and Consumption Patterns in Northern Belize and the North-Eastern Peten*

ISBN 978 1 4073 0909 5



BAR Publishing is the trading name of British Archaeological Reports (Oxford) Ltd. British Archaeological Reports was first incorporated in 1974 to publish the BAR Series, International and British. In 1992 Hadrian Books Ltd became part of the BAR group. This volume was originally published by Archaeopress in conjunction with British Archaeological Reports (Oxford) Ltd / Hadrian Books Ltd, the Series principal publisher, in 2012. This present volume is published by BAR Publishing, 2016.

Printed in England

BAR
PUBLISHING

BAR titles are available from:

BAR Publishing
122 Banbury Rd, Oxford, OX2 7BP, UK
EMAIL info@barpublishing.com
PHONE +44 (0)1865 310431
FAX +44 (0)1865 316916
www.barpublishing.com

## Abstract

Long-distance trade of obsidian in the Maya realm has been documented as early as the Middle Formative Period (1000-400 BC). Obsidian exchange continued in each succeeding period, through the Post-Classic (AD 900-1525), varying in both intensity and source of origin. It is the temporal variations in source utilisation that have formed the basis for obsidian research in the Maya area.

By focusing on the origin of the obsidian and the temporal context these studies provide valuable information in documenting shifts in source utilisation. However, by omitting the context of material recovered these studies have limited themselves to documenting these shifts. It is the intent of this work to demonstrate, by the presentation of a new research model, the importance of context as an analytical element in obsidian studies.

This work focuses on the distribution and consumption of obsidian on an intra-site scale with an intent to determine if variation in source utilisation can be attributed directly or indirectly to contextual variations. For the purposes of this study three types of context were identified, functional, archaeological and social. Information regarding obsidian consumption was complied from ten sites and two survey areas spanning four geographic regions across Northern Belize and North-eastern Peten. This data was first analysed using the standard time/source model to confirm the presence of temporal variations in the consumption of material from different obsidian sources, before more complex analyses integrating the various contexts into these patterns were performed.

It is through the creation of these latter paradigms that a comprehension of how, and by whom, obsidian from the various sources was being used obsidian may be ascertained. From these simple questions a broader perception of why obsidian was being used may be gained.

## TABLE OF CONTENTS

## LIST OF ABBREVIATIONS

| | |
|---|---|
| C | cache |
| CE:M | cutting edge to mass ratio |
| CF | construction fill |
| COL | collapse |
| E1 | elite 1 |
| E2 | elite 2 |
| EC | El Chayal Obsidian |
| EC | Early Classic |
| ELC | El Chayal Obsidian |
| FF | floor fill |
| FS | floor surface |
| IND | indeterminate social context |
| IX | Ixtepeque Obsidian |
| IXT | Ixtepeque Obsidian |
| H | humus |
| LC | Late Classic Period |
| LF | Late Formative Period |
| M | midden |
| MF | Middle Formative Period |
| MX | Mexican Obsidian |
| ND | no date |
| NE1 | non-elite 1 |
| NE2 | non-elite 2 |
| NE3 | non-elite 3 |
| OT | Obsidian other than El Chayal, San Martin Jilotepeque, Ixtepeque or Mexican |
| OTH | other deposits |
| PZ/S | plough-zone/surface |
| Q | quality |
| RC1 | ritual/ceremonial 1 |
| RC2 | ritual/ceremonial 2 |
| SM | San Martin Jilotepeque Obsidian |
| SMJ | San Martin Jilotepeque Obsidian |
| TS | tomb shaft |
| UNK | unknown archaeological context |
| WS | workshop |

## ACKNOWLEDGEMENTS

This work comprises the vast majority of the research I initially conducted and submitted to the Institute of Archaeology in 2000 as part of the requirements for my Doctorate of Philosphy now suitablely amended for publication. As this has been derived from my thesis there counsequently are countless people who contributed in immeasurable ways both large and small. The first people I would like to thank are Warwick Bray and Jose Oliver, my supervisors at the Institute of Archaeology. I couldn't have asked for better tutors. Their support played a major part in the formation of my research and theories.

My fellow archaeologists at the Maya Research Program and the Blue Creek Ruin where I worked from 1992 to 1998 are also deserving of credit. Dr. Thomas H. Guderjan, Program and Project Director, who provided me with the opportunity to work in Belize, and granted me permission to study the obsidian artefacts from the Blue Creek Ruin for my thesis. W. David Driver, my colleague, competitor, nemesis, and friend. Kim A. Cox deserves special mention as it was because of him that I became involved in obsidian studies.

My research also owes a great deal to Dr. Michael D. Glascock and the people at the Missouri Research Reactor who conducted the neutron activation analysis on the Blue Creek material. Dr. Glascock kindly allowed my material to be included in his NSF Grant (SBR 95-03035). Without his help and expert skill this research would have been impossible.

I would also like to thank Mike Halliwell and Stuart Laidlaw in the Photographic Department at the Institute of Archaeology. It is thanks to Mike's skill and endless patience that I have photos to include with this work.

I am also grateful to the members of the Department of Archaeology in Belize, both collectively and individually. I want them to know that I appreciate all the help they have given me over the years, granting permission for the Maya Research Program to export the necessary artefacts and generally helping expedite matters of export and other bureaucratic logistics.

Two people who, while not being archaeologists, have contributed as much to my thesis as my archaeological colleagues are Lorelei Friesen and Ian Day. These people are truly the 'unsung' heroes of my work. Ian has been my friend for more years than I think either of us wants to admit, and has always been there to help. Providing me with laughs and tea, soothing my nerves when my computer hated me, and catching it before I lobbed it out the window. My data files are the product of his many hours of labour, instruction and his 'magic formula'[1]. There is a reason I study stone tools, and if you are ever curious ask Ian. He knows the depths of my technological incompetence. Lorelei deserves praise beyond recognition. Not only did she read my entire thesis, including appendices and bibliography but she also edited it too!

I would also like to thank my parents, Kenneth and Dorothy Haines for their emotional and financial support and who regrettably did not live to see this published, my sister, Karen Haines, and my God-Parents, George and Rita Minoff without whose love and support I wouldn't have accomplished this work.

Thank you.

[1] (=IF(ISERR(VALUE(F1)), “” , VALUE(F1)) – Ask Ian, all I know is it works.

**Dedication**

This is for my Dad,
who only cared that I was happy,

and

for my friends Lorelei and Ian,
who were there in the beginning and the end,
and never lost faith along the way,
even when I did.

## INTRODUCTION

Ruined temples and hidden jungle cities have intrigued generations of adventurers and archaeologists. The desire to investigate such sites in Central America has been documented as far back as the late 18th century, when Josef Estacheria, the President of Guatemala's Royal Audiencia, sent expeditions to Palenque (Coe, 1992:74). While these early expeditions were unremarkable to the point that their reports were quickly lost, later endeavours were far more successful.

With the work of such scholars and adventurers as Charles Brasseur de Bourbourg, John Lloyd Stephen, and Alfred Maudsley, the last hundred years have seen incredible advances in the area of Maya studies. The deciphering of Maya writing, and the excavation and consolidation of Maya cities have lead to a greater understanding of the Maya social and political organisations, along with the ritual practices and economic activities. It is in the last area that the focus of this work lies, in the exchange and distribution of obsidian, a long-distance trade item from the Guatemalan Highlands and Central Mexican region.

Obsidian research is not new to Maya studies. Recognised as an 'exotic', or long-distance trade item, the recording and analysis of obsidian as a distinct artefact type has been *de rigueur* in Mesoamerica since the early 1970s (Graham *et al.* 1972; Jack *et al.* 1972; Kidder *et al.* 1978; Moholy-Nagy 1975; Willey 1972). Over the years, as advances in science have been made, the techniques used to analysis obsidian have also improved. Today, thanks to developments in nuclear technology, it is possible to identify chemical elements within obsidian and use the information to identify the original source for artefacts recovered at sites hundreds of kilometres away from obsidian deposits (Glascock 1994, 1998). Coupled with information gleaned from ceramic chronologies and other dating methods, archaeologists are now able to identify obsidian to both source and temporal period. Indeed, these studies have become the norm for Maya obsidian studies (Cobean 1991; Ford *et al.* 1997; Fowler *et al.* 1989; Hammond *et al.* 1984; Healy *et al.* 1984; McKillop 1989; McKillop *et al.* 1988; Mitchum 1989, 1991, 1994; Moholy-Nagy 1975, 1984, 1999; Moholy-Nagy and Nelson 1987; Nelson 1985; Nelson and Howard 1986; Nelson et al. 1977, 1978; Rice 1984; Rice et al. 1985). While identifying the different sources of obsidian used by the Maya and documenting changes in this pattern through time provides us with an understanding of *what* material was being used and *when*, it does little to aid in our understanding of *how* these sources were used and, perhaps more importantly, it fails to contribute to our understanding of *why* these sources were used or *why* variations in the consumption patterns for each source occurred.

The purpose of this work is to present a new model for obsidian studies, one that addresses the *how* and *why* of obsidian use. To this end, data regarding obsidian use at several sites were examined with emphasis being placed on the contexts of the material. It is my opinion that context, one of the basic components of archaeological research, is as valid and necessary in obsidian studies as in other areas of archaeological inquiry. It is only through this type of analysis that we can hope to acquire the necessary information on *who* was using obsidian and *for what purposes*. Without knowing the *who* and *what* of obsidian utilisation, understanding the *why* of obsidian reliance and exchange will always be beyond our reach.

This work examines the intra-site distribution of obsidian artefacts at 12 sites and two survey areas in four geographical regions to determine whether variations in source utilisation can be attributed directly or indirectly to consumption patterns resulting in contextual variations. For the purpose of this work, three types of context were considered: functional, archaeological, and social. A full description of each context is provided in Chapter 2, Section 1. Analysis of the material was conducted using three primary variables: obsidian source, temporal period, and context. Each context, functional, archaeological and social, was examined in turn. The results of these examinations will be discussed first in holistic terms before being divided into the smaller geographical areas to determine whether regional variations exist.

Although, advances in science have made it possible to identify obsidian to its source, it is possible that the Maya may not have known the original source and may rather have selected material based on more prosaic attributes or other external variables. In an effort to ensure that the focus on source utilisation did not obscure any non-source related patterns, the material is also discussed in terms of possible quality variations. These last investigations also include the variables time and context.

To aid in the presentation of this new model, the first section of Chapter 1 will provide background information regarding the geography of the region, cultural divisions, and basic chronology. The second section of this chapter will discuss resource diversity and trade among the different regions.

Methodological considerations and a background to the sites and survey area included in this study are explained in Chapter 2. The first of the two sections discusses the techniques used in analysing the material and defines the contextual terms used throughout the analytical process. This section also details the various techniques used in identifying the obsidian artefacts to their source of origin and provides information regarding the nature and types of obsidian sources used by the Maya. As obsidian from several sites was included in this analysis, the second section of this chapter provides background information on these sites and their obsidian collections. The first site discussed is Blue Creek Ruin, Orange Walk District, Belize. Obsidian from this site was analysed by myself for this work and provides the largest and best-documented collection. Information regarding material from other sites was gathered from articles and reports written about the respective sites. These sites are Tikal, Cerros, Cuello, Colha, Kichpanha, Nohmul, Becan, Ka'Kabish, El Mirador, Moho Caye, Northern River Lagoon, while the two survey areas are the Central Peten Lakes and the Tikal/Yaxha corridor. All sites, with one exception are from northern Belize or the north-eastern Department of Peten, Guatemala. Becan, the one exception, is located futher to the north in the Rio Bec Area in Quintanna Roo, Mexico.

Chapter 3 merges the methodology explained in Chapter 2, Section 1, with the data presented in Chapter 2, Section 2. This chapter is divided into three sections. The first examines the material by source and time period in the traditional model of obsidian studies and establishes that variability in source utilisation does occur between the temporal periods. Section 2 elaborates on this pattern by including the context of the artefacts. Contextual examinations commence with the broader functional categories before examining the finer nuances of the deposits by archaeological contexts. These are followed by an examination of the material by source, time, and social context. Section 3 examines the material based on weight and quality as opposed to 'per piece' source analyses.

Although each section in Chapter 3 includes its own summary, Chapter 4 integrates the diverse patterns into a holistic interpretation of the consumption and distribution patterns. It is this chapter that delves into the questions of *who*, *what,* and *why* of obsidian patterns.

# CHAPTER 1
# INTRODUCTION TO THE MAYA REALM

## SECTION 1
## BACKGROUND AND CHRONOLOGY

### GEOGRAPHY OF THE REGION

The Maya area has traditionally been divided into two large resource zones, referred to as the Highlands and the Lowlands. These zones are based primarily upon elevation and vegetation. The Maya Highlands, found in the southern part of the Maya realm, are defined geographically as those areas above 500-600 metres (Borhegyi 1965a, 1965b), or culturally, separating the lowland linguistic groups (Yucatec, Putun, Palencano Chol, Manche Chol, Mopan, Chorti, Tojolabal, Tzeltal, and Tzotzil), from the Highland Quichen group (Thompson 1990). Vegetation in the Maya Highlands consists of a mixed evergreen/deciduous forest cover. Types of trees include oaks, sweetgum, dogwood, and a variety of pines, with pine predominating in areas of higher altitude (Sharer 1994:30).

Geologically, the Maya Lowlands consists of low limestone and karst hills in the south which open out to a wide, low limestone shelf to the north that forms the largest part of the Yucatan Peninsula (Sharer 1994:19-43). Two distinct biomes are present in the Maya Lowlands, the high canopy of the Central or Peten Region in the south, and the lower xerophytic vegetation of the Northern Yucatan and Coastal areas (Sharer 1994:33-34; Romney 1959:204-232). While divisions between these areas are often gradual and include transition zones, in places such as the Maya Mountains and, to a lesser extent, the Rio Bravo Escarpment in north-western Belize, the change can be quite dramatic.

### CULTURAL DIVISIONS

Traditionally the Maya Lowlands have been divided into ten area: Peten, or Central Area, Usumacinta River, Pasion River, Western, Southern, Puuc, Chenes, Rio Bec, Northern Plain of the Yucatan and Coastal Belize/East Coast (Appendix I, figure 1). While originally designed to demarcate areas of distinct architectural style (Pollock 1965), these zones are now defined more broadly, and include ceramic assemblages as part of their defining characteristics. Although often drawn with clear boundaries, the reality is that these regions possess no absolute borders and the lines drawn on maps tend to conform more to topographic phenomena than to archaeologically distinct divisions, and the merging of ceramic and architectural traits along these border areas is not uncommon.

The Peten, Western, Southern, Usumacinta River, Pasion River and the Coastal Belize/East Coast areas form the Southern Maya Lowland. Both the Usumacinta and Pasion River areas are quite small, scarcely extending beyond their respective watershed basins. The Western Area is slightly larger, consisting roughly of the lower Usumacinta River basin to Laguna de Terminos in the east, and the Grijaval River basin to the west. The largest of the Southern Lowland zones is the Central Peten area. This area encompasses much of the central Yucatan peninsula, and was once thought to be the heartland of Maya cultural development (Rathje 1972). While today early settlements have been discovered in many parts of the Maya Lowlands (Adams 1981; Awe 1990; Awe and Healy 1995; Freidel 1977; Fowler *et al.* 1989; Garber and Reilly III 1995; Haines 1996; Haines and Suthers 1997; Haines and Wilhelmy 1999; Hammond 1977, 1991, 1995; Hammond *et al.* 1991, 1995; Hansen 1992; Rice 1976; Sullivan 1991; Vail 1988), to date the earliest major ceremonial centres, El Mirador and Nakbe, have been found in the Central Peten (Hansen 1990, 1992; Howell 1989; Matheny, Hansen and Gurr 1980; Sharer 1994). Several early sites have been found in the Coastal Belize area (Freidel 1977, 1978; Hammond 1976; Pendergast 1982, 1990), although none are on the same scale as those in the Central Peten. The Coastal Belize area encompasses much of the country, starting at the off-shore cayes in the east and terminating at the Rio Bravo Escarpment in the north and the Maya Mountains in the south.

The Northern Lowlands is composed of the Puuc, Chenes, Northern Yucatan, and Rio Bec areas. Only one site from these four regions, Becan in the Rio Bec area, was included for analysis, as it may have been served by the same river trade route as the other sites examined. The Rio Bec area is a geological and environmental transition zone between the high canopy and rugged river-valley terrain of the south and the drier, lower bush vegetation and flat riverless environment in the north. Sites in this area are distinguished primarily on the basis of architectural styles. Distinctive to this area are structures designed with pyramid-like towers, steep non-functional stairways, and solid superstructures with false doorways (Pollock 1965:427-428).

Sites included for analysis in this report are located primarily along the Rios Hondo and Azul. Additional sites were included that, while not directly on these rivers, were considered to have been within the potential scope of trade networks using these river routes. As such, the analysis area encompasses aspects of Coastal Belize, Rio Bec, and Central Peten areas.

### CHRONOLOGY

Maya history has traditionally been divided into three main periods prosaically referred to as the Formative,

Classic, and Post-Classic Periods. As these periods tend to span considerable lengths of time, the Formative Period lasting over 2000 years, over the years they have been subdivided into smaller classifications, each as imaginatively named as the larger unit within which they are contained. These terms have evolved and been refined as new information, particularly regarding the Formative Period, has been discovered. Material examined in this report spans the Late Formative through to the end of the Late Classic, a period sometimes referred to as the Terminal Classic. As continuity of occupation was deemed an important consideration, and many of the sites occupied in the Late Formative and Early Classic suffered population decline or abandonment by the end of the Late Classic, Post-Classic material was not considered. Hence, only those periods directly related to this study will be outlined here.

**FORMATIVE PERIOD (2000 BC – AD 250)**
In American literature, this period is commonly referred to as the Pre-Classic period. The term 'Pre-Classic' was based on dividing Maya cultural development into temporal periods on the presence or absence of traits externally perceived as being indicative of a 'classic civilisation'. This term was defined more than 30 years ago, and since then considerable archaeological work has been accomplished, with the result of adding significantly to our understanding of early Maya culture. The term 'Formative', deriving from the idea of a 'formation period', has been suggested as an alternative designation for the time preceding the Classic Period. As it is my opinion that it more accurately defines the events of this period, this latter phrase is used in this report. The Formative period is the longest of the three periods, and is divided into three smaller phases, the Early, Middle and Late Formative Periods.

**Early Formative Period (2000 BC – 1000 BC)**
Very little is known about the Early Formative Period. Spanning roughly one thousand years, from 2000 BC to 1000 BC, its existence is based more on speculation than fact. As material remains from the Middle Formative Period reveal a level of sophistication well above that normally associated with a culture recently transformed from an archaic, or pre-ceramic, society to a settled, ceramic society, it has been presumed that transitions from archaic to formative development must have occurred considerably earlier, despite the lack of current evidence. It is believed that both ceramics and settled village life began during this period. Structures are believed to have been simply pole and thatch buildings similar to those still in use in many parts of the Yucatan today.

**Middle Formative Period (1000 BC – 400 BC)**
Using a new chronology put forward by Sharer (1994), the Middle Formative Period begins around 1000 BC and lasts until approximately 400 BC. Domestic structures were elaborated slightly from the preceding period to include a roughly squared stone retaining wall, an indication of greater investments of manpower. Burials also show greater investment along with diversity of artefacts included as mortuary items. Individuals were buried with an assortment of items, ranging from nothing to carved bone beads, jade, and elaborate vessels. This period is also marked by a widening of trade networks. Obsidian from the Highlands and jade are found for the first time, and marine shells fashioned into bracelets show contact with the coast.

Ceramics from this period represent the first complete complex known from the Lowland Maya region. The ceramics of this complex, known as Mamom, demonstrate a high degree of homogeneity throughout the southern Maya Lowlands. They have been seen as lacking specialised types for discrete functions – such as burials and caches – and it is generally considered that people were interred with the same types of vessels that would have been available to them on an everyday basis (Sharer 1994). Although this may be true in general, in recent years vessels have been discovered in caches and burials that may belie this original assumption (Williams-Beck 1997).

It is during the Middle Formative period that we find evidence of early, integrated communities. One of the earliest settled sites is Cuello in northern Belize. Occupation at this site began around 1200 BC. Recent work at the sites of Nakbe and El Mirador in the Central Peten suggest that Nakbe was a settled village by 1300 BC, and underwent a transformation to a more complex society with monumental architecture between 600 BC and 400 BC (Hansen 1992:60-61). The site of El Mirador lagged only slightly behind its neighbour, with monumental architecture being constructed around 300 BC (Hansen 1990).

**Late Formative Period (400 BC – AD 250)**
Although once it was considered quite rare, now few sites are found that do not contain Late Formative material. Ceramics from this period belong to the Chicannel Sphere, a clear development from Mamom. These vessels show a widespread standardisation of form, finish, and decoration, and are monochromatic with reds to reddish browns and oranges being the dominant colour.

One of the most famous sites from this period is Cerros, a relatively small settlement located at Chetumal Bay in Northern Belize. This otherwise minor site has received much attention over the years due to the presence of a series of zoomorphic masks located across the front of a temple dating to roughly 300 BC (Freidel 1986b). These images reveal a complex cosmology, incorporating the patterns of various celestial bodies, deities, and ritual symbols. The consensus is that this monument is as an iconographic representation of the will of the cosmos and the divine rulership of an elite group (Freidel and Schele 1988; Garber and Reilly 1995; Schele and Freidel 1990). It is considered one of the earliest and best pieces of evidence for the existence of a stratified ruling elite, and many of the images present on the structure continued to be used through the Classic period as signs of rulership.

The end of the Late Formative Period, dating from roughly 50 BC to AD 250, is occasionally referred to as either the Proto-Classic or Terminal Formative Period. This period is often seen as representing a dramatic change in the material culture of the Lowland Maya. James Gifford, in his defining work on the ceramics from Barton Ramie, was a major proponent of this period (Gifford 1976). These changes are considered by such proponents to represent the influx of culturally different (although not necessarily distinct) groups into the lowlands. However, these changes are not present in the rest of the material culture, and the presence of vessels of this type are limited, appearing mostly in burials and deposits from the upper stratum of the society. Furthermore, excavations have yielded little other evidence of changes that cannot be attributed to a general, and natural, intensification of cultural activity. James Brady and others argue rather successfully that the term "Proto-Classic" has little validity beyond explanations of ceramic chronologies (Brady *et al.* 1998). As such, this term was not used in this report, and material dating to this period was considered Late Formative.

Evidence from the Formative Period has revealed that many of the institutions and attributes previously believed to define the Classic Period are either present or developing in this earlier period.. These include monumental public architecture, complex ideology, hierarchical social organisation with elite rulership, and long distance trade. The only major characteristic from the Classic Period that has not been discovered in the Formative Period is writing. Even this last attribute was probably present in the Formative Period. As the Maya word for stela means "tree" it is possible that the early carvings were made on wooden monuments that have not survived the centuries.

## CLASSIC PERIOD (AD 250 – AD 900)

This period was once seen as the time of florescence for the Maya culture. It was defined originally as the period in which monumental architecture, religious and social complexity, and the long count calendar were in use (Morley 1946). We now know that, with the exception of the long count calendar, virtually all the attributes previously used to define this period may be found in the preceding Formative Period (Freidel and Schele 1988; Hansen 1990, 1992; Schele and Freidel 1990; Sharer 1994). The Classic period is divided into three, and sometimes four, smaller temporal units. The two most common are the Early Classic (AD 250 – AD 600) and the Late Classic (AD 600 – AD 750/900). Two other periods that are often used are the Middle or Intermediate Classic (c. AD 500 – AD 700), and the Terminal Classic (AD 750 – AD 900). Of these periods the latter, the Terminal Classic, is more commonly used. The former, also referred to as 'the hiatus' (Adams 1991:196-197; Thompson 1965:343) and Late Classic I (Maxwell 1996:46), refers to a point in Tikal's history when the construction of monumental structures and the erection of stelae ceased (Coggins 1979; Mohoy-Nagy 1994, 1999). As this phenomenon is restricted to Tikal and does not noticeably effect the archaeological record of any of the other sites analysed it was not included in this study.

### Early Classic Period (AD 250 – AD 600)

Spanning AD 250 to AD 600, the Early Classic Period is marked by increased inter-site political activity. The presence of elites and *ahauob*[2] (rulers) becomes institutionalised. The solidification of this ruling class is accomplished through a complex series of political, economic, and ideological routes. Imagery used to identify *ahau* and authority become codified and relatively standardised throughout the Maya Lowlands. These images include manikin sceptres, Kawil images, the double-headed serpent bar, and the regalia of *ahauob* which creates the impression of them as a living version of the sacred World Tree (Schele and Freidel 1990).

The political activity of the Central Lowlands appears to have been dominated by Tikal in northern Guatemala. This site appears to have possessed an almost precocious social development, rapidly outstripping neighbouring sites in terms of architectural construction and political expansion. Early in this period, Tikal begins expanding its sphere of political influence and absorbing other sites, starting with Uaxactun roughly 40 km to the north. A definite realignment of power structures occurred during this period, indicated by the decline or abandonment of many Formative period centres, including Cerros and El Mirador.

The presence of stelae recording the existence and activities of rulers during this period allows us to reconstruct the dynastic histories of many sites. These monuments also provide valuable information regarding inter-site marriages, political alliances, and ritual activities. Information from stelae, coupled with that from archaeological excavations, make this period one of the best known in the Maya realm.

Monumental architecture becomes a standard feature at sites and the core areas of many sites were initiated during this period. The main acropolis at Tikal was initiated at this time, as were major constructions at Seibal, Piedras Negras, Yaxchilan, Altun Ha, Lamanai, Caracol, Calakmul, and Palenque to name but a few sites. Sites became more elaborate with central precincts possessing

---

[2] The suffix '*-ob*' is used in Yucatean and Chol to indicate a plural word (Schele and Freidel 1990:21).

large plazas, tall temples, and in many cases the beginnings of acropoleis.

Ceramics also become more elaborate, and polychrome vessels not seen in the previous period emerge. These ceramics demonstrate radical stylistic shifts from those in the previous period. Surface slips become more glossy, losing the 'waxy' feel characteristic of Mamom wares, and the predominant colour becomes orange, in place of the preceding red. Decorations are typically red and black painting on orange or cream base with motifs executed in bands and repetitive geometric patterns. Distinctive attributes of this period include medial or basal flanges and ring bases.

Trade between the different regions flourished during the Early Classic. Large quantities of jade, obsidian, and other non-local items are found in burials and caches dating to this period. It has been suggested that sites co-operated to control access and distribution of resources both on a local and inter-regional scale (Arnaud 1990; Brown 1977; Zeitlin 1982).

The end of the Early Classic period is marked by the collapse of Tikal's political authority. Evidence suggests that the defeat of Tikal was the result of a co-operative alliance between Caracol, Calakmul and several other smaller surrounding sites.

**Late Classic Period (AD 600 – AD 750)**
With the downfall of Tikal, a political vacuum appeared in the Central Lowlands. Several sites which during the Early Classic period had been small polities begin to develop rapidly, expanding and filling this vacuum with a series of competitive cities, all striving to succeed Tikal, actively contending for control. Sites that participated in this power struggle included Caracol and Calakmul (the protagonists in Tikal's defeat), Dos Pilas, Yaxchilan, and Palenque. Alliances between sites become common in Late Classic politics, and there are numerous mentions of women from ranking lineages marrying into dynasties at other sites (Schele and Mathews 1991; Sharer 1994). However, there is no evidence that the lowlands were ever unified either politically or economically during this period. Rather, the political landscape consisted of a patchwork of co-operating, autonomous polities of different sizes and strengths.

Architectural variations become more pronounced during this period, giving rise to distinctive regional styles. Ceramics also become more elaborate and highly decorated. The geometric patterns of the Early Classic give way to 'codex'-style paintings. With their images of elites and supernatural deities engaged in various activities, these vessels provide an interesting look into Late Classic Maya practices and ideologies.

This period is perhaps best noted for its precarious political and social nature. The competitive activities of the various polities, coupled with an increasing population and subsequent environmental exploitation, formed a volatile mix.

**Terminal Classic Period (AD 750 – AD 900)**
Originally classed simply as part of the Late Classic, the period between AD 750 and AD 900 has become known as the Terminal Classic. This period is marked by the intensification and eventual collapse of the various Late Classic phenomena. Competition turned to violent warfare, exploitation of the environment led to declines in productivity, and social stresses resulted in shifts in political structures leading to a dissolution of the authority of rulers and initiating a period of power-sharing among elites.

Architectural construction in the southern lowlands declines, as does the erection of monuments, although areas in the Northern Yucatan around Chichen Itza continued to flourish. The last stela erected using the Long Count date was in AD 909 at Tonina. Trade in long-distance status items ceases in the southern lowlands by the end of the Terminal Classic, as do the cultural achievements of the previous periods.

Although not standardised in Maya Lowlands studies, researched on several sites referenced in this work make use of this temporal designation. As the designation reflects an important period of political realignment, it was deemed important to include this period in the obsidian analysis conducted in this report.

At the end of the Terminal Classic Period, many sites were either abandoned or suffered a severe social and political restructuring. In many cases, the exact nature of the organisation of the individual polities is unclear. Consequently, it was decided to limit the scope of this work to those periods outlined above. Information gleaned about the obsidian consumption and distribution patterns during these periods may contribute to and, it is hoped, clarify issues of intra-site economic activity.

## SECTION 2
## TRADE AND EXCHANGE AMONG THE MAYA

### INTRODUCTION

That "man doth not live by bread alone" (Deuteronomy viii:3) is as true for the Maya as it was for the ancient populations of the bible and most peoples since. While maize is generally considered to be the primary subsistence staple of the Maya, their diet included other cultigens such as beans, chillies, and varieties of root crops among other plants. Maya comestibles were not limited to such crops; both terrestrial and marine fauna contributed to the list of available consumables. However, goods needed or desired by Maya communities and the individuals therein were not limited to the basics of food production, nor were they constrained by everyday subsistence requirements. Perhaps it is a basic factor of human behaviour to want, and therefore to try to obtain, what is different, exotic or simply rare even if a locally available item or material would serve equally well.

The Maya were no different in this behaviour, and many sites yielded evidence of the acquisition of artefacts manufactured from non-local material, even when some if not all of these artefacts could have been manufactured from material that was available locally. Granite from both the Maya Mountains and the Highlands for use as manos and metates may have been a sign of status as its exclusive appearance in elite contexts at Blue Creek indicate (Pastrana, personal communication); locally available silicified limestone while perhaps not as hard nor as durable as granite, would also serve for grinding corn and would have been easier and cheaper to obtain. Likewise, although obsidian is undeniably sharper than chert, tools manufactured from cryptocrystalline silicates will still cut and scrape hides, and this material was better suited for heavier, agricultural tasks (Andrefsky 1998). In truth there was little that is crucial for basic subsistence that is not available locally in the Maya lowlands (Marcus 1983), with perhaps the exception of salt (Marcus 1983, Andrews, 1984, MacKinnon and Kepecs 1989).

### GEOGRAPHIC AND RESOURCE DIVERSITY

Although the majority, if not entirety, of material required for daily subsistence was available locally, many items desired and used by the Maya were available only from a distance. This diversity in resources is a reflection of the diversity of topography and vegetation discussed previously. While raw material and goods produced from non-local materials formed the basis of Maya trade, it has been speculated that the Maya exchanged ideas along with material goods (Rathje 1978; Freidel 1979).

The Maya realm once encompassed virtually the entire Yucatan Peninsula, from the Caribbean Sea to the Pacific Ocean. Today this territory is divided among several countries, and Maya sites may be found in parts of southern Mexico, Belize, Guatemala, and in the western sections of El Salvador and Honduras. The topography of the area is highly diverse (Appendix I, figure 2), ranging from volcanically active metamorphic mountains in the south that rise over 3000 metres, through low rolling limestone and karst hills as one moves north, then out onto the broad, low limestone shelf that, at less than 200 metres above sea level, forms the majority of the Yucatan Peninsula (Sharer 1994:19-43). This natural diversity has led to the Maya world being divided into regions referred to as the Highlands and the Lowlands.

Diversity of topography is naturally accompanied by a diversity in vegetation and natural resources. Vegetation in the Highlands consists of a mixed evergreen- deciduous forest cover. Types of trees include oak, sweetgum, dogwood, and a variety of pines, with pine predominating in areas of higher altitude (Sharer 1994:30). The Lowlands may be subdivided into two distinct areas based on vegetation: the high canopy of the Central or Peten Region, and the lower xerophytic vegetation of the Northern Yucatan and Coastal areas (Sharer 1994:33-34; Romney 1959:204-232). While divisions between the Highland and Lowland areas are often gradual and include transition zones, some areas, such as the Maya Mountains and, to a lesser extent, the Northern Escarpment in Belize, show quite dramatic divisions. Another noticeable and significant division within the Lowlands is the presence of numerous rivers in the hilly southern and central sections and the virtual absence of any major rivers in the northern limestone shelf. Rio Candelaria on the west coast and the Rio Hondo on the east coast form the northernmost boundary of this lacustrine environment.

### TRADE, COMMODITIES, ROUTES, AND MECHANISMS

Marcus suggests that the "long-distance" trade should not be viewed as a means of securing the basic supplies needed for existence, but as a type of "foreign relation" (Marcus 1983: 479). Drennan also notes the unfeasibility of transporting commodities, subsistence or otherwise, over distances greater than 275 kilometres (Drennan 1984a, 1984b). However, a more pertinent question is raised by Andrews (1984), who notes that the problem with discussions of trade is rooted in their failure to adequately define "long-distance" versus "local" trade. Andrews suggests that an "exchange between polities that are several hundred kilometres distant" would "have to suffice" for a definition of 'long distance trade' (Andrews 1984:827). However, Sidrys in his analysis on exchange systems suggests a more tangible terminology, where local trade is defined as exchange within a 50 kilometre area, and long-distance trade as exchanges between communities 400 to 500 kilometres apart (Sidrys 1983a). Sidrys also includes a intermediate area which he identifies as a regional trade zone, extending 50 to 400 kilometres.

Although still extremely broad, Sidrys' attempt to create a more refined system of exchange areas is useful as it allows for variations in trade mechanisms as well as the nature of exchanged commodities, to be explored. The Maya are known to have traded a variety of commodities both within and between the different geographic areas (coastal, arboreal, mountain). Rathje (1972) proposed a model of cultural development among the Maya based on the need of the Central Peten for resources. Although this theory, based on the trade of material goods in exchange for dissemination of knowledge for social development, has been effectively disproved in the wake of the discovery of Middle Formative deposits and *in situ* social development at numerous sites both within and beyond the Peten, the evidence of non-local trade goods in early deposits reveals the existence of trade networks as early as the Middle Formative Period. Although these networks undoubtedly underwent organisational changes though the centuries, trade of non-local items on both a local level (within 50 kilometres) and inter-regionally has been documented for both the Formative and Classic Periods.

While the most enduring trade objects are obviously inorganic items such as ceramic wares and stone objects, a variety of items are known to have been traded throughout the Maya realm. Perishable utilitarian goods such as honey, salt, cacao, cotton, terrestrial and marine animals, among others items were undoubtedly traded on a local level as well as between regions. Ritual, non-utilitarian or status objects such as jade objects, stingray spines, *Spondylus* shells, and shell beads were also traded between geographic areas and appear at countless sites throughout the Maya Lowlands (Garber 1983, 1989; Buttles 1992; Cobos 1994; Maxwell 1994, 1996; Guderjan 1995, 1998; Lee and Awe 1995; Haines 1996, 1999; Haines and Wilhelmy 1999). Obsidian, which appears to have served both utilitarian and non-utilitarian functions, is also found at many sites throughout the Lowlands (Andrews *et al.* 1989; Asaro *et al.* 1978; Awe 1994; Dreiss 1986; Dreiss and Brown 1989; Ford *et al.* 1997; Fowler *et al.* 1989; Graham *et al.* 1972; Guderjan *et al.* 1989; Hammond 1976, 1989; Hammond *et al.* 1984; Healy *et al.* 1984; Hester and Michel 1980; Jackson and Love 1991; Johnson 1976; McKillop 1989, 1995; Moholy-Nagy 1984; Moholy-Nagy and Nelson 1990, 1987; Neivens and Libby 1976; Nelson *et al.* 1977, 1983; Nelson and Howard 1986; Rice 1984; Santley *et al.* 1986; Rovner 1989; Spence 1996).

A full discussion of the nature of trade is beyond the scope of this work. Furthermore, the exact nature or organisation of Maya trade systems is unclear. Various models have been put forward over the years. The concept of centralised redistribution centres focusing on Tikal and Kaminaljuyu appeared frequently in the literature from the early 1970s to 1990 (Rathje 1972; Sanders 1973:353, 1977; Brown 1977; Nelson 1985; Sidrys and Kimberlin 1979; Arnauld 1990). This centralised system has been challenged by theories suggesting exchange revolved around trading stations (McKillop 1995, 1996; Mitchum 1994:14) and ports of trade (Guderjan *et al.* 1989; Guderjan 1993, 1995a; Andrews 1990; McKillop 1996), and others that see trade not as an economic function but as a socio-politically integrating method between area and regional polities (Tourtellot and Sabloff 1972; Marcus 1983; Drennan 1984a, 1984b). While a completely different rationale for the role of trade in Mesoamerican cultures suggests that it was a function in the creation and maintenance of stratification and elite status (Chapman 1971; Parsons and Price 1971; Freidel 1981, 1982).

Although, merchants or "agents of trade" (Rathje *et al.* 1978:148) like their modern Maya counterparts (Hammond 1978) probably carried a variety of goods serving equally varied purposes, this work focuses on the consumption of obsidian at an intra-site level. Consequently, it follows that any discourse on trade should focus less on the esoteric and debated nature of trade as a whole in the Maya realm and concentrate chiefly on the transportation and exchange of the material in question. Several papers have been written discussing the routes and nature of obsidian exchange (Hammond 1972; Sanders 1973, 1977; Sidrys and Kimberlin 1979; Arnaud 1990; Mitchum 1994; McKillop 1996).

One of the first concerted efforts to understand the networks of obsidian exchange was presented by Hammond in his seminal article *Obsidian Trade Routes* (1972). In this work Hammond, while acknowledging contemporaneous utilisation of the El Chayal and Ixtepeque sources, attributes the distribution of these materials to separate trade routes – the El Chayal material being exchanged via an overland route based on the Chixoy-Pasion Rivers and the Ixtepeque material being transported along the Motagua River and the Caribbean coast. The exclusivity of these routes appear to have been based on ideas of convenience and are not supported by either McKillop's 1996 study nor by research presented by this work. However, as the Maya possessed no 'beasts of burden' but relied upon human carriers or possibly dogs to transport goods from one region to another, the use and importance of the rivers in the Maya trade networks has become an accepted fact (Rathje 1972; Hammond 1976; Johnson 1976; Sidrys 1976; Adams 1978; Edwards 1978; Rathje *et al.* 1978; Vail 1987; Guderjan 1988, 1993; Chase and Chase 1989; McKillop and Jackson 1989; Guderjan and Garber 1995). Both Adams (1978) and Tourtellot (1978) conducted experiments in transporting goods via river routes. The consensus was that while the movement of goods via river routes allowed for the transportation of larger cargoes per individual, there was no substantive decrease in transportation time (Adams 1978; Tourtellot 1978). Increases in speed by river travel were relative to the direction and strength of the current, and what savings may have been made travelling downriver could be expected to have been commensurately expended on the return trip. Consequently, inter-regional trade has been argued to have been limited to small groups of individuals, and most likely elite in occupation or limited to elite status items (Adams 1978; Drennan 1984a). Yet inter-regional trade not only existed, but judging by the volume of obsidian discovered at some of the Lowland sites, it was a flourishing practice in the Early and Late Classic periods (Neivens 1976; Pendergast 1981; Rice 1984; Lewenstein 1987; Moholy-Nagy 1994).

Material from the El Chayal source comprises the majority of pieces recovered from Classic Period contexts, with Ixtepeque, San Martin Jilotepeque, and even material from Central Mexican sources rounding out many collections. McKillop (1996) noted an association between sample size and number and proportions of sources represented at coastal sites. For Late Classic sites where more than 10 artefacts were tested, the results showed 83 per cent of the collection consisted of El Chayal material, while 13 per cent were Ixtepeque in origin and four per cent were from the San Martin Jilotepeque source. In cases where fewer than 10 artefacts were tested, the results showed a marked increase in the proportion of El Chayal material, which constituted 94 per cent of the collections, with Ixtepeque obsidian accounting for the remaining six per cent. San Martin Jilotepeque material was not identified as being present in these smaller samples. This tendency should be remembered when considering smaller samples, and is a strong incentive for larger and more representational test collections. The tendency towards one source over another may also be influenced by the nature of the sample being tested. Evidence presented in this work reveals a correlation between context and source (see Chapter 3 this volume), indicating that representational samples need not only be proportionate to the overall collection, but equally distributed among the recovery contexts. An example would be the large quantities of Pachuca obsidian contained within the Post-Interment Offering associated with Tomb F-8/1, an elite burials at Altun Ha (Pendergast 1990:266-267; Spence 1996). In this case, the green obsidian accounted for 115 pieces of the 182-plus pieces, roughly 63 per cent of the collection (Pendergast 1990:266-267). If this was the only context for which obsidian at Altun Ha was tested, it would reveal an inaccurate pattern suggesting a greater than actual consumption of Pachuca and Central Mexican obsidian at this site than is normally seen in the Maya Lowlands (see Chapter 3 this volume).

McKillop argues that the Maya obsidian trade, and Maya economy in total, functioned along similar lines as modern society, where individuals and communities responded to "the forces of supply and demand, based on the concepts of rational choice, scarcity, and maximisation" (1996:52). I would suggest, based on the evidence presented in this volume (Chapter 3) and elsewhere (Pendergast 1990; Spence 1996; Moholy-Nagy 1989, 1999) that the decisions made regarding the sources of obsidian utilised had less to do with supply than with conscious choice. The presence of larger than average quantities of specific obsidian sources in certain contexts (Pendergast 1990; Spence 1996; Chapter 3 this volume), and the paucity of other sources in specific contexts (Chapter 3 this volume), suggest that obsidian was being chosen based on criteria other than supply or availability. The following chapters present a new model of obsidian analysis with the intention of creating new avenues of investigation that may assist in our understanding of Maya obsidian trade, and perhaps further our understanding of Maya economy.

## CHAPTER 2
## OUTLINE OF ANALYSIS

### INTRODUCTION

This study examines intra-site obsidian distribution patterns. The purpose of this examination is to identify potential contextual and social variations in material and source utilisation through time. Changs in the patterns of obsidian utilisation will then be examined to determine whether these variations may be linked to changes in social organisation or political structure of Maya communities.

The analysis will focus on three primary aspects: 1) material, 2) time period or date, and 3) context. The material will be examined initially in regards to original source; however, discussions as to the quality and quantity of the material will also be included. Establishing temporal contexts for the material will allow changes in source utilisation and archaeological and social contexts to be documented through subsequent periods of occupation. By identifying the contexts of the obsidian, consumption patterns for the material may be understood. This last analytical attribute is the crux of this report, for while previous site reports document sources by time period and any attendent shifts in utilisation of sources through time, the inclusion of context into this rather formulaic analysis, while advocated (Arnaud 1990; Moholy-Nagy 1989), is rarely applied (Dreiss *et al.* 1993; Moholy-Nagy 1989, 1999).

Source identification was conducted at the two sites examined by this author, Blue Creek and Ka'Kabish, Orange Walk District, Belize, through a combination of neutron activation analysis (NAA) and visual identification. NAA testing was conducted by Michael Glascock at the Missouri University Research Reactor (MURR). Visual identification was carried out by the author following criteria identified by the author and Kim Cox. Identification of material from other sites included in this study was conducted and published by their respective excavators. While NAA was the most common technique used in these reports, x-ray fluorescence was also employed.

Temporal identification is crucial if the patterns, and changes in such, are to be tracked through the development and life of a community. Material from the sites examined by this author were dated primarily by ceramic chronology, although some deposits were dated by radio-carbon testing. Information regarding dates for material from other sites was taken as reported in their associated articles. Often, the method of acquisition of these dates was not specified. In an effort to standardise the material into a usable format, traditional time periods, such as Late Formative and Early Classic, were employed.

Examination of the possible contexts for the artefacts resulted in a dual system of classification. One type of context established was excavation context; the second was social context. Excavation context identifies the tangible location of the material when discovered (i.e. construction fill, floor surfaces, caches, etc.). Examination of this type of context is intended to illustrate changes in where or how obsidian was used. Social context is more problematic, as it relies upon identification of the probable status of the individuals involved in the creation or use of the excavation context. The intention of this context is to document possible changes in who, or what segments of Maya communities, were consuming obsidian at different points in time. Both types of contexts may be used to illuminate changes in ideologies within Maya communities.

While the Maya traded a large variety of items (Thompson 1967, 1970; Graham 1988), artefacts fashioned from obsidian are singled out for this study. Obsidian was chosen for several reasons: 1) it is clearly of non-local origin, 2) it appears in both utilitarian and ritual context, 3) it is present in varying degrees during all relevant stages of Maya history, 4) it is easily sourced to its original geological location, 5) preservation is high, and 6) its recorded presence at numerous sites allows for comparisons to be made between different areas.

Obsidian, a volcanic glass found in the Guatemalan Highlands and Central Mexico roughly 450 km and 1200 km respectively from the study area, first appears in the archaeological record of Maya sites in the Middle Formative Period (1000 BC – 400 BC). At this time the material is limited in quantity and is distributed relatively evenly throughout sites (Spence 1982). During the later Early Classic Period (AD 250 AD – 600 BC), the material begins appearing in greater quantities, while at the same time appearing in fewer types of deposits. It is also during this period that a distinctive shift in source utilisation from San Martin Jilotepeque to El Chayal obsidian initiated in the Late Formative Period (Awe and Healy 1994; Dreiss and Brown 1989; Fowler *et al.* 1989; Nelson and Howard 1986; Nelson *et al.* 1983; Rovner 1989) becomes widespread throughout the Maya region (Dreiss and Brown 1989; Hammond 1991b; Hammond *et al.* 1984; Moholy-Nagy1984; Moholy-Nagy and Nelson 1987, 1990; Nelson *et al.* 1983; Rice 1984; Rice *et al.* 1985). A second more gradual shift from El Chayal to Ixtepeque obsidian begins in the Late Classic Period with the introduction of material from this latter source (Dreiss *et al* 1993; Hammond *et al.* 1984; McKillop 1995; McKillop *et al.* 1988; Smith and McField 1996). It is these changes and how they appear in the archaeological record that is the focus of this study.

## Section 1
## Analysis Techniques

### Methodology

Reports from 10 sites and two survey areas were found in the archaeological literature with sufficient information regarding their obsidian samples to warrant inclusion in this study. These sites are Tikal (Mock 1997; Moholy-Nagy 1984; Moholy-Nagy and Nelson 1987, 1990), Cerros (Mitchum 1989, 1994), Cuello (Hammond 1991a, 1991b), Colha (Dreiss *et al.* 1993), Kichpanha (Dreiss *et al.* 1993), Nohmul (Hammond 1983; Hammond *et al.* 1984), Becan (Rovner 1989), El Mirador (Dreiss *et al.* 1993; Nelson and Howard 1986), Moho Caye, Northern River Lagoon (Dreiss *et al.* 1993; Mock 1997). The two survey areas are the Central Peten Lakes (Rice 1984; Rice *et al.* 1985), and the Tikal/Yaxha Corridor (Ford *et al.* 1997). Two other sites, Blue Creek and Ka'Kabish, were investigated by the author and provide the framework for this study. These sites are grouped into four geographical regions: 1) coastal, 2) river oriented, 3) inland, and 4) Central Area/Peten. A full discussion of these groups and the sites may be found in Section 2 of this chapter.

In order to understand the changing patterns of obsidian distribution and consumption, various aspects of distribution and consumption were considered. The three most important variables were material, time period or date, and context of the obsidian. Numerous reports detail the different sources utilised by various sites through the Maya Formative and Classic periods (Awe and Healy 1994; Dreiss and Brown 1989; Dreiss *et al.* 1993; Ford *et al.* 1997; Guderjan *et al.* 1989; Hammond 1976; Hammond *et al.* 1984; Healy *et al.* 1984; Heizer *et al.* 1965, 1971a, 1917b; Johnson 1992; McKillop 1989; McKillop *et al.* 1988; Mitchum 1994; Moholy-Nagy 1975, 1976; Moholy-Nagy and Nelson 1987, 1990; Rice 1984; Rice *et al.* 1985). Differences in source utilisation during the various Maya periods has become a recognised, if not predictable, factor when discussing changes in source emphasis, the most notable of these being the shift away from San Martin Jilotepeque material in Guatmala, and an increased reliance upon El Chayal material, a source to the south of San Martin Jilotepeque, during the Late Formative period, culminating with its dominance in the Early Classic . Reasons for this shift in utilisation of sources through time is often attributed to political pressures in the source production area (Dreiss *et al.* 1993; Zeitlin 1982), the most common being the influence of Kaminaljuyu in promoting the El Chayal source and 'shutting down' production of the San Martin Jilotepeque source (Brown 1977; Nelson 1985; Sidrys and Kimberlin 1979; Michel 1976). However, very few reports consider the end consumer in their analysis, nor do they provide information regarding the final, intra-site destination of the material. This study examines the possibility that changes in sources were the result of an increased demand for specific materials induced by ritual activities and the subsequent creation of an elite hierarchy.

Consequently, the foremost factor considered in this analysis was the archaeological context in which the material was discovered, and what this implies about the systemic context in which it initially functioned. While it is a general principle in archaeology that "artefact type, approximate date, and archaeological context should always be published, so that correlation between these attributes and the source of the analysed material can be sought" (Moholy-Nagy and Nelson 1990:75), most articles discussing obsidian finds report only two of these criteria, source and time period being the two most common, with some attempt to differentiate between artefact types. The few that do include context do so in such a way as to require an almost Herculean effort to wring the information from the text. A full discussion of the types and nature of the contexts used follows this section.

The primary concern of this paper is the creation and application of a viable model to further the study of intra-site obsidian distribution and consumption patterns. In an effort to make this model applicable on a wider, inter-site level, the criteria used in this analysis were constrained to those variables currently recorded in the archaeological literature, although other factors were considered and some were suggested as being potentially more useful.

As the predominant form of reporting obsidian objects is per piece, this is the format used for the larger inter-site geographical and regional analyses. However, in-depth examination of the material from Blue Creek and Ka'Kabish included analysing distribution of the material based on weight and source quality. These criteria were checked against the context of the artefacts (excavation, social, and temporal) to determine whether either quantity-by-weight, or differences in quality influenced the distribution pattern of the obsidian within these communities. Use-wear analysis was considered, but was found to be beyond the scope of this report. However, note was taken of the difference in the severity of edge-damage in relation to the excavation contexts to determine possible avenues of future research.

Analysis of material from the other sites was conducted in two stages: the material was first included in a holistic data base, then assessed on the basis of its geographic group. The reasons for this two-stage analysis were to investigate the possibility of different patterns of obsidian intra-site distribution resulting from sites' positions along the river route, and to identify the variables which contributed to the creation of the distribution patterns. As the largest and most contextually diverse collection came from Blue Creek (n=1,124), this material was also analysed separately to help identify potential varibles for the distribution of obsidian.

## WEIGHT VS. PIECE

As data regarding the material from sites other than Blue Creek and Ka'Kabish was collected from previously published reports, the details were restricted to the prescribed format of the information given. The standard method of presentation for obsidian artefacts is by the piece. However, this method is not without its attendant problems. The most notable of these appears when attempting to discuss access to resources. Using the current standard of referring to artefact quantities by raw numbers, six obsidian blades found in a domestic structure are equivalent to six blades placed in a cache. However, if the six blades in the first group weigh an average of 0.5 grams each, and the blades in the second group weigh on average 1.0 grams each, then stating the quantity of resources consumed by the deposits by their raw numbers would lead to the assumption that the two events consumed equal quantities of obsidian. This assumption is patently incorrect, as the cache can be clearly seen, using weight as the comparative variable, to have used twice as much obsidian as the household. However, from a functionalist perspective, it might be argued that six blades will function as six blades regardless of their weight and that it is the utility of the object that needs to be considered. Consequently, we run into an almost Zen-like parable of 'do six blades always equal six blades?'.

Sheets and Muto (1972) suggest that a better indicator of the quantity of material consumed at any locale is a ratio calculated on the length of cutting edge (CE) to the total mass of the blade (M). Higher cutting edge to mass ratios (CE:M) are then taken as an indicator of conservation of the material: the higher the ratio the scarcer, and potentially more valuable, the material (1972). Calculations of the total cutting edge were made by measuring the maximum length of the blade, from striking platform to furthest distal point, and multiplying by two to account for both edges. Sheets and Muto extended this calculation to include the arrises present on the surface of the blades, multiplying the previous calculated length of the blade by the number of arrises, as they felt these flake scars would have been used for scraping.

However, this approach, like others before it, comes with inherent difficulties. The most notable of these is that thickness of a blade may have a negative effect on the ratio, making two blades each with a ratio of 74 mm to 1 gram equal, although one blade may be a very fine blade 20 mm wide and only 3 mm thick, while the other blade may be a scant 8 mm wide and 5 mm thick. The difference in these blades is not readily apparent unless one understands the process by which blades are constructed. A full description of the technique was detailed by Don Crabtree (1968), and it is to his groundbreaking work that I would refer the reader for in-depth discussion. In brief, the process of obsidian blade construction involves the blades being removed from a polyhedral core in a circular pattern. As the blades are removed, the core becomes smaller, resulting in the blades becoming narrower, and more distinctly curved. This process continues until the core is exhausted. Consequently, size of blades may be the result of their location on the core. If sites farther from obsidian sources, or further down a trade route, consistently possess higher CE:M ratios, the differences in ratios between locations may not represent differential status, but simply a case of initial access to the material.

Furthermore, as Braswell, Andres, and Glascock (1994:185) noted in their analysis of material at the sites of Copan and Quelepa, the CE:M ratios may be misleading for inter-site comparisons. Analysis of obsidian from the elite residential zone at Copan, a larger and "more important site than [its] contemporary Quelepa", revealed that these blades had a "significantly greater" CE:M ratio. According to Sheets and Muto (1972), this pattern would indicate a higher level of conservation at the Copan residences. This is a situation that Braswell *et al.* feel is highly unlikely, and I concur based not only on the social status of the individuals in residence, but also on the proximity of Copan to obsidian resources. Braswell *et al.* reason that "the artisans who produced the prismatic blades used at Quelepa *could not* or did not choose to make blades as fine as those produced and used in contemporary Copan." (Braswell *et al.* 1994:185, emphasis added). That the obsidian knappers at Quelepa, a smaller and less prestigious site *could not* make blades of the same fineness as the craftsmen at Copan is an important consideration. Quelepa is a smaller and less prestigious site than Copan and may not have had access to the same level of craftsmen as Copan. Blades produced by less skilled, or apprentice, craftsmen may have been cruder and therefore possessing a higher CE:M ratio.

Although this theory has a few flaws, it is not without applicability in obsidian studies. As the CE:M ratio analysis is a material-economising strategy, it may be more relevant for examinations of intra-site consumption patterns than for comparisons of inter-regional patterns (Sidrys 1979). Sidrys' use of this analysis technique relies on the assumption that areas farthest from obsidian sources would exhibit higher degrees of material conservation. Although at first glance this theory may seem quite reasonable, as the CE:M ratios for regions consisted of averages for all sites available within those areas, it fails to account for the possibility of redistributive centres. Furthermore, by considering material holistically for each site, possible variations between areas or moieties within communities are masked. It is to identify intra-site patterns that CE:M ratios are advocated here.

Applying the same 'conservation' rationale, used by Sidrys on a smaller, intra-site scale, one may hypothesise that areas with restricted access to obsidian would exhibit higher CE:M ratios. This technique may be of value in determining relative status of various individuals residing in different areas of a site.

Analysis of the Blue Creek and Ka'Kabish material included an examination of the CE:M ratio technique. However, the process used to obtain the required measurements differed from that of Sheets and Muto by two significant means. The first difference was that, unlike the previous study, measurements of each cutting edge were taken separately. These lengths were calculated, not from the striking platform, but from the 'shoulder' of the blade, then added together to provide the total cutting edge (CE) measure. It is the opinion of this author that measurements taken from the shoulder, as opposed to the striking platform, represent a more accurate assessment of used blade edge, eliminating the 'heel' of the blade that would have been difficult if not impossible to use for cutting or scraping. This measurement was recorded using jewellery callipers to one tenth of a millimetre.

The second major difference was that arrises were not included in the calculation. Arrises were excluded from this study, as an examination of the collection revealed that many of the blades possessed irregular, or non-continuous patterns of flake scars. This factor would make any multiplication of blade length by the number of arrises higher than the probable true value. Furthermore, utilisation of these flake scars is indeterminate without an in-depth analysis into use-wear and use-scratch morphology (Hay 1977). Previous studies in this field (Lewenstein 1981, 1987; Hay 1977), have revealed that various tasks utilising the edge of the blades leave scratches and polish across the entire surface of the blade, making identification of arris use versus edge use virtually indeterminable. The CE:M variables collected for the Blue Creek and Ka'Kabish materials were checked against the three primary aspects of time, context and source, to ensure that no possible attributes that may affect the patterns of distribution and consumption were overlooked.

### Questions of Quality

The problems discussed above in regards to differences in blade size lead to the question of identifying potential differences in quality of blade manufacture. Quality can be, and usually is, a very subjective criterion. Attempts to distinguish a *high quality* blade lead to discussions of craftsmanship, and, to use a rather well known cliché, 'beauty is in the eye of the beholder'. While it should be recognised that differences in thickness and length may be indicative of varying qualities of craftsmanship, to attempt to assess how the Maya may have rated these pieces is dangerous territory, far beyond the scope of this paper.

However, assessments on the quality of the source material may be attempted. Experiments in lithic reproductions have shown that obsidian "must be devoid of inclusions, grain, or undetected flaws, for the slightest imperfection will hinge-off the blades and render the core useless for further blade removal" (Crabtree 1968:451). Crabtree elaborates further and notes that even the smallest of flaws may prevent blade production through multiple problems, including collapsing the pressure platform and causing the flow of pressure to terminate early before the blade is completely detached (Crabtree 1968:452). Either event would result in few blades from any given core and greater post-production waste (Crabtree 1968; Sheets and Muto 1972). As the quality of the obsidian has been seen to have a direct effect on the quality and quantity of the blades, the possibility exists that this was a factor in determining which material was distributed to the various segments within the community. Objects were examined under a ten powered magnifying glass with evaluations made based on the effect ofinclusions on the production of the blade. Quality was graded in three categories based on inclusions visible in the surface texture. Material containing no, or few, tiny rocky particles (specks requiring a 10x loupe and careful examination to identify), and possessing a smooth glassy surface, resulting in straight even blade edges and arrises, was considered to be the finest grade. Blades with fine rock particles resulting in a pitted surface, or with bands or layers creating variations and irregularities in the surface texture, were considered to be second grade material. Objects with numbers of large inclusions, and irregular surfaces and edges, were considered to have been constructed of lowest grade material (figure 2.1).

### . Use-Wear And Edge –Damage

While not officially a part of this analysis, a word regarding use-wear analysis is nonetheless warranted. Several studies have been conducted with the attempt to determine an object's use based on discernible wear patterns While the majority of these works focus on chert tools, some studies have been conducted on the use-wear patterns on obsidian tools (Huracombe 1994; Lewenstein 1981, 1987; Lewenstein and Walker 1984; Hay 1977). Studies on obsidian tools have shown that for analysis to have any viable results, the tools are best examined under a high-powered, scanning-electron microscope and compared with patterns created on modern reproductions of the tools in question (Hay 1977; Lewenstein 1981, 1987). While use-wear patterns may be distinguished on the modern reproductions, identification of the patterns on the original objects is more difficult, as, unlike the reproductions, there is no certainty that the Maya used different tools for different tasks. Indeed, many tools analysed appeared to have a variation of patterning (Lewenstein 1981). Furthermore, Hurcombe identified four different processes -- Manufacture, Use, Excavation Method and Post-Excavation Processing -- through which obsidian may acquire

Photograph Showing the Various Qualitites of Obsidian
Left, Quality 1; Middle, Quality 2; Right, Quality 3

Figure 2.1

unintentional, or accidental damage that can be mistaken for evidence of use (1992:72-78). A fifth means whereby obsidian may acquire accidental damage was identified during analysis of the Blue Creek material, this was the nature of the archaeological deposit from which items were recovered. Not surprisingly, pieces with a high degree of damage came from construction fill contexts (Appendix III), and locations from the site settlement zone that had been cleared and ploughed by the local Mennonite community. Because of the potential for ambiguous results, and since it is of only peripheral interest to the focus of this study, in-depth analysis of possible function of the Blue Creek obsidian tools through study of their wear patterns was left for later study.

## ARTEFACT TYPE

While the majority of obsidian artefacts recovered are fragments of prismatic blades, other types of obsidian artefacts are known from the Maya area (Kidder *et al.* 1978; Moholy-Nagy 1994; Willey 1972). These other artefact types include flakes, expended cores, and bifaces. Despite the tendency of reports to discuss obsidian per piece, it was felt that the different types of obsidian artefacts should be noted. Obsidian artefact types identified among the Blue Creek and Ka'Kabish collections are listed below along with a brief definition of the characteristics of each type. Note was taken of the types of artefacts appearing in the different types of deposits.

### Cores

Prismatic obsidian blades are removed from polyhedral cores (Crabtree 1968). While the method by which obsidian blades were acquired by a community has been debated, the most commonly accepted theory is that obsidian was shipped in the shape of a pre-formed nodule from which blades could be struck off at the final destination (Moholy-Nagy 1994:68). A method which dispenses with the worry of breakage en route. The presence of expended cores in various deposits supports the idea that at least some of the material was shipped in this manner. Although all the cores discovered at Blue Creek were used, some of them appear to have been fully expended during blade production, while others appear to have fractured during the manufacturing process. Several of these fractured cores appear to have been used as expedient tools. As a result, five types of cores were identified: 1) cores that, while utilised, retain a proximal surface, 2) cores that are fully expended and have no visible proximal surface, 3) fragments of cores created through either the inadvertent fracturing of a core during blade production or deliberately for rejuvenation of the core, 4) fractured cores that were utilised as an expedient tool, and 5) sections of cores, or core tips, that were removed to rejuvenate the core and increase the number of blades obtained from that core.

### Blades

This category comprises the majority of artefacts found. Prismatic blades are struck from polyhedral cores, and are longer than they are wide or thick (Crabtree 1968). Due to the technique used in blade production, these blades often possess arrises along their ventral side, giving the blades a triangular or rhomboid form in cross-section. Those pieces identified as 'blade' are intact artefacts, with clearly identifiable proximal and distal ends. Other sub-categories of blades include blade fragments (broken medial sections) and broken proximal and distal ends (Appendix I, figure 6).

### Flakes

Flakes are often difficult to differentiate from the broken, proximal end of a blade if the broken piece is small enough. Identification is complicated further by the fact that most 'flakes' appear to be miss-struck, or failed attempts to produce a blade. However, careful observation of the distal end of the piece should enable an accurate identification of the object. Flakes are wider than they are long or thick. Although use-wear analysis is not a part of this study, note was taken if the flakes appeared to have been debitage, or to have been used as expedient tools.

### Bifaces

These are bifacially worked tools, and include both lanceolate or lenticular blades (Appendix I, figure 7), and small, notched points. Although these two types of artefacts are sometimes placed in separate categories, it was felt that for this study they would best grouped together. As obsidian is very fragile and production of bifaces would require more effort and material, it was deemed that, for this study, these artefacts had enough in common to justify grouping them together. Only two collections included in this analysis that contained bifaces, or pieces of bifaces. These materials came from El Mirador and Blue Creek.

### Unifaces

This is a type of tool that has only been worked on one side. Several of the larger chunks show signs of having been used as uniface tools. The majority of these appear to be informal scrapers (Appendix I, figure 8).

### Chunks

These are amorphous pieces of obsidian that cannot be identified as belonging to a blade, flake, or core. While these pieces are generally small fragments, under one centimetre, they can occasionally be larger. When the larger pieces were used as expedient or formal tools, they were noted as such under the headings of biface or uniface.

**Decorations**
While chipped around the edge into circular shapes, these pieces are too small to have served as tools (Appendix I, figure 9). Unlike the "tinklers" found at Kaminaljuyu (Kidder *et al.* 1978), these small, flat pieces were not drilled in the centre. Similar objects were discovered at Tikal where they were described as inlays (Moholy-Nagy 1994:69).

## CHRONOLOGICAL FRAMEWORK

Maya cultural development is traditionally divided into three main chronological periods: the Formative or Pre-Classic Period, the Classic Period, and the Post-Classic Period, with each being further subdivided into early, middle, and late, or terminal, facets. As these time periods are defined by the presence or absence of certain cultural attributes, the absolute (calendrical) chronology is often debated, although the range of variability between authors is minimal. A full discussion of the each period and their characteristics may be found in Chapter 1, Section 1.

Although several of the papers used in this research include the Proto-Classic Period as one of the temporal divisions, the dates generally provided for the Formative and Classic Periods are closer to those found in Sharer's recent work (1994), than in the more traditionally accepted Belizean chronology provided by Gifford (1976). Consequently, for the purpose of this work Sharer's chronology has been adapted to include the Proto-Classic Period, and employs the following periods:

| | | |
|---|---|---|
| Classic -- | Terminal | AD 750 - 900 |
| | Late | AD 600 - 750 |
| | Early | AD 250 - 600 |
| Formative -- | Late | 400 BC - AD 250 |
| | Middle | 1000 - 400 BC |
| | Early | 2000 - 1000 BC |

It should be noted that the Post -Classic Period, while mentioned in several of the obsidian reports from other sites, is not included here. The focus of this paper is on the Formative Periods through to the end of the Terminal Classic Period. The following abbreviations were used to identify the various chronological time periods in which the artefacts were recovered: Middle Formative (MF), Late Formative (LF), Early Classic (EC), Late Classic (LC), and Terminal Classic (TC). Assignment to the narrowest possible temporal period was preferred and used whenever possible.

## CONTEXTS

Archaeological study is primarily the analysis of a culture's material remains, leading to inferences about the abstract characteristics of the society. The material remains and the abstract aspects of the culture examined are then used to reconstruct the society as a whole. As material in the archaeological record is *systemically static*, in that it is no longer functioning in behavioural patterns, we must examine the context of the artefact to determine its possible role in the society.

Schiffer (1972, 1976,1977) identifies two types of artefact context, *archaeological* and *systemic*. Archaeological context is defined as the "non-behavioural state of cultural materials", while systemic context includes those "materials within an ongoing behavioural system -- [things that] are handled or observed" (Schiffer 1976, 1977). Using these two contexts, Schiffer identifies four types of cultural transformations which artefacts can undergo. These cultural transformations are: 1) from systemic to archaeological context (S-A), 2) from archaeological to systemic context (A-S), 3) from state to state within the archaeological record (A-A), and 4) through successive systemic states (S-S) (Schiffer 1976, 1977). Artefacts that fall into the first category include items that were discarded, lost, abandoned, or interred with a burial. In each of these cases, material that was once involved in the behavioural pattern of the community has been removed. The second category of cultural transformation, A-S, entails objects being reintroduced into the behavioural system, either through scavenging, if they are discrete objects, or mass integration, in the case of refuse deposits reused as construction fill. Transformations that occur within the archaeological record (A-A), often involve upward or downward migration of artefacts resulting from surface disturbances, ploughing, discing, alternations of freezing and thawing processes, and erosion. The last category, successive system transformations (S-S), includes processes of recycling, secondary use, and conservation (Schiffer 1976, Schiffer 1977).

While these categories have been used successfully in some analyses (Lewenstein 1987), they deal with very general principles. Upon scrutiny, they cannot always explain the processes through which an item has entered the archaeological record. An example of this is caches, which Schiffer classes as items that have undergone a systemic to archaeological transformation (1976:33). While it is undeniable that these materials are no longer circulating among the community members, to suggest that they are no longer systemically participating implies an understanding of the ideology involved in the creation of the cache. One suggestion given to explain the act of caching is that it serves as a means of removing material from circulation for reasons of economics. Weiner has suggested that instead of removing objects from a community, caches served as a means of ensuring that objects would *remain* with a lineage or community (Weiner 1985, Weiner 1992). These objects were identified as *inalienable possessions*, objects which through association, veneration, and caching activities would always be identified with and belong to the group. They are not 'lost', 'abandoned' or 'discarded', nor are they 'disposed of' along with a deceased individual. In the case of Maya caches, it has been suggested that their incorporation into a building lends identity to the structure (Becker 1992). As the Maya placed these caches in almost identical locations in subsequent construction episodes of the same structures, it seems ludicrous to suggest that succeeding generations were unaware of the previous offerings. Consequently, one might reasonably argue that, while the caches were beyond being "handled or observed" (Schiffer 1976:28, 1977:16), they were still participating in the behavioural patterns of the society. It is the opinion of this author that these items, along with other items in a similar situation discussed below, should be considered to have experienced a transformation of systemic contexts.

While Schiffer's four types of cultural transformations are useful in analysing the possible means by which material moved through and eventually out of the behavioural system of the community, these categories, as noted briefly above, are not without their problems. In the case of my analysis, the most pressing of these problems is the gross lumping of various types of discrete behavioural patterns. As this study is designed, in part, to analyse the consumption and distribution patterns of obsidian artefacts, differentiating between whether an artefact was lost, discarded, or deliberately included in a burial or offering is vital. It is not sufficient to state that the artefact in question was moved from the systemic to the archaeological context without clarifying the intention in the transfer. Consequently, more specific categories detailing the nature of the deposits will be used as the divisions for contextual analysis, while an adaptation of Schiffer's analysis of cultural transformations will be considered in discussing the movement of materials between the systemic and archaeological records.

Garber, in analysing the artefacts from Cerros, created a typology of contexts by selecting descriptions that were easily identified in the excavation record (Garber 1989). Some contexts (burials, caches, habitation debris, construction fill, etc.) were based on associated features, while others, whose associations were unclear (surface, humus, fall, slump, etc.), were identified on the basis of the excavation matrix or provenience in which they were found. Moholy-Nagy (1997), in her discussion of various types of deposits at Tikal, also argues that certain deposits may best be identified on the basis of the excavated context. In her work, Moholy-Nagy uses the term *recovery context* to refer to "the archaeological setting in which the materials occurred" (1997:299).

While agreeing with the principle of use terms based on recognisable definitions in the archaeological literature, I prefer the term *excavation context* when referring to categories of excavated objects. The word *excavation* brings with it an almost ingrained, or predetermined, set of terminology, one based in the archaeological literature, and including such categories as midden, construction fill and cache. Conversely, the word *recovery* conveys a broader definition of any location where something may be found, whether that be a burial, backdirt pile, rubbish bin or the crisper drawer of a refrigerator.

While a contextual typology similar to that of Garber's (1989) or Moholy-Nagy's (1997) was created for the analysis conducted here, it was felt that such an analysis was limited in its interpretative value. Contexts used by previous authors, based on archaeological or excavation criteria, do not differentiate between an item found on the floor of a temple or of a domestic residence. A floor, under their criteria, is simply a floor. However, as this study is interested in differential consumption patterns, it was felt that distinctions should be made in regards to the functional and social environment. Consequently, two additional contextual tiers were added in this report. The idea of 'functional' contexts is a simplification of Schiffer's idea of c-transforms combined with the concept of excavation contexts. In this model the material is grouped according to perceived nature of the deposit. From this model patterns that might otherwise be obscured in the more detailed examinations of the minute nature of each excavation context may be highlighted. As such this model may be considered a broader analysis of the excavation, or archaeological contexts. The second set of contexts is based on locating the deposits within a social hierarchy. The purpose of this additional contextual level is to discern differential access patterns that may have existed between different social strata.

Both types of contexts, excavation and social, are listed below. The terms used in each category are listed alphabetically and are accompanied by a discussion of the term and a definition of its use in this report.

### EXCAVATION CONTEXTS

Contextual types included in this category were deliberately selected to represent the spectrum of identifiable deposits included in archaeological reports. Some of these contexts are discrete individual features (i.e. caches), while others are identified by the excavated fill in the surrounding unit (i.e. humus). By creating a typology that encompasses both types of deposits, features and fill, it is felt that any ambiguities in the contexts can be accommodated.

Terms considered in this classification include Cache, Construction Fill, Collapse, Floor Fill, Floor Surface, Humus, Midden, Mortuary Good, Tomb Shaft, Plough-Zone or Surface, Special Deposit, Workshops and a catch-all category listed simply as Other. Using an adaptation of Schiffer's broader categories of cultural transformations, the terms derived for this study can be grouped into one of five categories. Four of these categories are similar to Schiffer's original terms: deliberately discarded (S-A), reused (A-S), conserved (S-S), displaced (A-A). The fifth category that has been added here is for objects from unknown or uncertain contexts. This category is unfortunately often necessary in practice as the final position of materials recovered sometimes defies accurate definition. Such objects are often displaced from their original contexts in such a way as to preclude their consideration as having undergone an archaeological-to-archaeological contextual transformation.

Three of the contexts described below, Construction Fill, Collapse, and Humus, fall neatly into the category of material that has undergone an archaeological to archaeological transformation (A-A). This material has been displaced from its original position and, in the case of the Construction Fill or Collapse, reused elsewhere in the community. It is this concept of deliberate repositioning that is the criterion for this category. Material that, while displaced, does not fit into this category, includes objects found in the ploughed fields surrounding the site, material from the humus level, and items whose excavated context for one reason or another is unclear. Complete definitions for these groups are listed below under the titles of Humus, Plough-Zone or Surface, and Other.

Objects found in the following deposits are considered indicators of behaviour designed to remove objects from the systemic system: Floor Fill, Floor Surface, Midden, and Workshop. While Schiffer would also classify artefacts in the categories of Cache, Mortuary Good, Tomb Shaft or Special Deposits as disposed of items (Schiffer 1977:33), it is the opinion of this author that while these objects have indisputably been removed from easy access, it is debatable that they no longer fulfil a systemic function. Consequently, this report considers these objects to have undergone a transformation of systemic contexts (S-S) and will be considered objects that have been 'curated', as opposed to 'lost' (Schiffer 1972, Schiffer 1976, Schiffer 1977). No objects found could be identified as having re-entered the systemic record. Consequently, this category of objects will not be discussed.

**Cache (C)**

This category is widely used in archaeological literature. Derived from the French word *cacher*, meaning "to hide" or "to conceal" (Webster 1950), this term is used in archaeology to refer to "an artefact or group of artefacts intentionally placed in a specific location unrelated to a burial" (Loten and Pendergast 1984:5).

Several different types of caching behaviour have been identified from Maya ritual deposits. The two most prominent types are 'dedicatory' and 'votive' or 'offertory' (Coe 1959, 1965a; Maxwell 1996; Schele and Freidel 1990; Smith 1950). Dedicatory caches are considered to be those deposits that were placed in a structure during construction as an integral part of the building process. Votive, or offertory, caches consist of material that was interred in the structure after the completion of the building, probably associated with ritual activity conducted at the structure. Walker (1995) has elaborated upon these definitions, categorising cache material as belonging to one of three types, 'sacrificial', 'krataphoneous' or 'ceremonial trash'. Walker's new classifications have merit, particularly in the realisation that not all caches, and the materials contained therein, are participating in an event, but may be the result of the disposal of by-products from an event. However, his classification implies an understanding of the intention behind the caching action that I am doubtful can always be ascertained, specifically when the deposit contains a variety of artefacts from stingray spines to jade beads (Walker 1995). The former artefact may reasonably be argued to be considered ceremonial trash having likely been used in a bloodletting ritual, however, the jade beads are more problematic as their function beyond a status symbol or their economic value as an indicator of wealth has not been fully investigated.

Moholy-Nagy feels that the act of caching was "associated almost exclusively with the elite" (1997:302); however, excavations at the site of Blue Creek have revealed that this activity was more widespread. The primary difference between caches found in public, or elite, contexts and those in non-elite residences appears to be the quantity and quality of items interred. Non-elite caches usually contain a limited number and type of items, with non-local artefacts restricted to a few small jade beads (Clagette 1997, n.d.). Where multiple items were discovered, they tended towards ceramic vessels with no small offerings of either local or non-local material (Clagett n.d.).

Elite or public caches exhibit greater diversity in the type and material of the offerings. Items found in these caches included a variety of marine material (Driver 1995; Maxwell 1994, 1996, 1998; Willey 1972), obsidian blades (Coe 1990, 1988; Moholy-Nagy 1994; Guderjan 1995; Haines and Wilhelmy 1999; Pendergast 1986, 1998; Smith 1972; Weiss 1995; Willey 1972), jade beads and plaques (Gilgan 1996; Guderjan 1996; Guderjan and Weiss 1995), specular hematite (Gilgan 1996), mercury (Pendergast 1986), and osteological remains, both human and faunal in origin (Cheetham 1994; Coe 1959, 1990; Haines 1996; Maxwell 1996, 1998).

Although Loten and Pendergast agree that "the artefacts that comprise a cache were presumably intended as an offering", they argue that "the term 'cache' is preferable because it is a designator without functional implication" (1984:5). This is a difficult concept to accept as the very nature of the word cache implies a function, namely the secreting or storing of material. Consequently, this report defines the term as the deliberate interment of objects distinct from a burial, and, while outside the general systemic realm of portable material culture, not without possible ideological or social function.

**Collapse (COL)**
This category has been previously defined as "material that has resulted from decay of a structure since its abandonment" (Loten and Pendergast 1984:5). However, this description is a little vague in that it does not clearly address the issue of material found in ambiguous contexts, such as items possibly belonging originally in construction fill, or on floors, but no longer clearly identified as such. Often, as a result of excavation strategies, these items are classed as having a mixed deposit, collapse/room fill and humus. As the fill that comprises walls and roofs of structures is generally more carefully selected than that of substructures or platforms, this category was created to differentiate artefacts found associated with a structure from those that appear in the general ballast of a larger construction project, and may be in secondary or tertiary context (see Construction Fill below).

**Construction Fill (CF)**
In the case of larger construction projects, such as substructure platforms, it is not unusual to incorporate material from different contexts into the fill. One such common incorporation is material from middens and other no-longer-necessary deposits (see definition below). The use, or reuse, of such materials serves a twin function: first, it neatly disposes of unsightly and unwanted material, and second, it provides ballast for a structure with less effort than quarrying raw material. However, as pleasant a prospect as this practice may have been to the Maya, it presents certain problems for archaeologist, as it makes identification of the primary systemic origin for objects found within the fill of a structure virtually impossible. Consequently, rather than attempt a contextual definition on the original deposit, a process that would be based on almost pure conjecture, objects found in this context are defined by excavation matrix.

Materials considered to be construction fill are those deposits that compose the bulk of the construction for a structure, either in the form of the substructure or the walls and basal platforms. These deposits are sometimes referred to as *ballast*, "a layer or bed of inert material, such as stone, installed as a base for a floor" (Loten and Pendergast 1984:4), or, in the case of larger works, as *core*, "internal or hearting masonry" (Loten and Pendergast 1984:6). Objects found loose in the construction fill (i.e., not contained in caches or tombs), are considered to have been from recycled deposits and therefore the systemic context is indeterminate. Consequently, for the purpose of this study, these items are not assigned a social or functional context beyond indeterminate (IND).

**Floor Fill (FF)**
Material that was found in the aggregate beneath or between floors was classed separately from those items found in general construction fill. The reason for this separation is the nature and origin of the fill. Material used in the final preparation for the laying of a plaster floor, or the replastering of a previous surface, is generally referred to as 'aggregate', an "inert material that is mixed with binder of mortar, clay, or other material to form concrete" (Loten and Pendergast 1984:3). Aggregate, unlike the larger ballast used in core construction, is often laid with more care and "usually consists of quarried stone fragments, nodules, or rubble, ranging from pea to fist size" (Loten and Pendergast 1984:3). The smaller fragments are necessary to ensure a smooth surface that will not sag or buckle in time. Small portions of ceramics and other debitage are frequently found in this level, either material from reused middens or, in the case of replastering of a previous floor, possibly material from the surface of the previous floor. As the latter case would reflect the material being utilised by the residents of the structure, it was deemed useful to distinguish this material from that of the larger construction core. The different deposits, floor fill versus construction fill, are also distinguishable in terms of their larger role in the behaviour of the community. Items found in the general construction fill, as discussed above, most probably in a secondary or tertiary context, having been moved from one archaeological context to another. Conversely, items beneath a floor may reflect a pattern of lost or primary discard context.

**Floor Surface (FS)**
This category is fairly self explanatory. Artefacts classed under this heading were found in close association with a floor, either within a few centimetres of the surface or in the softened plaster matrix. This scope was deemed necessary to account for possible post-depositional movement and degradation of plaster surfaces. Objects found on these surfaces are seen to indicate pattern of loss or abandonment.

**Humus (H)**
This category is strictly an excavation category with no associated cultural meaning. Material is grouped under this heading based on the associated matrix in which it was found. Humus is generally considered to be

"decomposed organic matter" (Whitehouse 1983:222), and in humid rainforest environments comprises the top several centimetres of an excavation.

Artefacts recovered from this excavation context are considered to be from indeterminate contexts, both archaeological and social, due to the severe activity by roots, as well as insects and other creatures within this layer. Consequently, they are not discussed beyond that of the matrix from which they were recovered. This context should not be confused with that of *Plough-Zone/Surface*, discussed below.

**Midden (M)**

Waste products leave the systemic system through many avenues, as has been discussed above. The term midden is used among archaeologists to refer to "a concentration of cultural debris" (Bahn 1992), without differentiating the contents of such dumps. Generally this term refers to mixed deposits of organic and non-organic material, and "the material in which the debris is encapsulated" (Bahn 1992). Further discussions have delved into the different types of waste and waste management (Hayden and Cannon 1983; Moholy-Nagy 1997). Work has also been done to differentiate between types of waste, and has consequently refined the definitions of refuse, debitage and garbage to refer to discarded durable material, objects created during production of artefacts, and biodegradable waste, respectively (Moholy-Nagy 1997). While these distinctions are valid, they are not, with perhaps the exception of debitage, widely used in the archaeological literature. More often, the terms waste, garbage, and refuse appear to be interchangeable, and where one term is used consistently, it is often not specifically defined.

To further complicate this situation, disposed material may have been moved from the place where it was initially used and placed in specific areas (middens). Material found where it was originally deposited is considered to be in primary context. This is considered quite rare, as by their very nature the majority of objects disposed of are moved to another location for disposal. An example of this would be a broken obsidian blade. The blade may have broken during use within a house but, as obsidian is quite sharp, it is unlikely that the pieces would have been left on the floor, in their primary context, but would instead be moved to an out-of-the-way location, a secondary context (Clark 1991:72; Lee Decker 1994). In the Maya area this material may also be found beneath later constructions where it may have been redistributed to form an even base for a floor or platform, and sometimes may have been incorporated into larger construction projects as fill (Haviland 1981; Coe 1990:878; Moholy-Nagy 1997), forming a tertiary context.

Consequently, the term *midden* will be used here to indicate any layers of occupational or cultural debris found in primary or secondary context. As this definition is fairly general, where the term appears in reports by other researchers, without a clear definition, it was assumed to possess the same definition. One exception to this definition is deposits composed solely, or almost exclusively, of stone debitage. These latter deposits are referred to in some of the reports used as *workshops* (Hester *et al.* 1983; Hester and Shafer 1984, 1991), although the term *workshop refuse* would be more accurate. As the definition of *workshop* has a clearly associated function separate from that of general refuse, and is the term widely used in previous literature, it was decided to retain this segmentation of terms.

Material found in construction fill is considered in the category of *construction fill*. The reason for this is that the material in question is considered to have undergone what Schiffer defines as a transformation of state without having left the archaeological record (Schiffer 1972, 1976, 1977). These pieces are considered to have been viewed not as individual objects, but in a group with the midden matrix as recyclable material.

**Mortuary Goods (MG)**

Artefacts classed under this heading include only those objects directly associated with, and possibly belonging to or used by, the individual. As goods found in the shaft of the tomb, above the capstone, or incorporated into the floor surface cannot be securely assigned an owner or origin within the society, they are classed separately (see below). No attempt was made to try to differentiate between goods included as *life possessions* versus those that were possibly *prestations* (Mauss, 1994). As all the goods, regardless of intention of interment, indicate access by that section of the society to various resources, it was deemed unnecessary to differentiate between the possible nature of the goods. However, as it is possible to assign social status to individuals based on the location of the burial, the mortuary goods will fall into one of four social contexts: Elite 1, Elite 2, Non-Elite 1, or Non-Elite 2. A full discussion of these social contexts can be found below.

**Tomb Shaft (TS)**

While noted at a number of sites (Haines 1995, n.d.; Moholy-Nagy 1994; Guderjan 1991; Trik 1963; Smith 1950, 1972), tomb shafts have not received much discussion in the archaeological literature. Although the term *tomb shaft* properly refers to the entire channel created during the construction of a burial chamber, it is used here to indicate the layers of non-ballast material used during the filling of a tomb shaft.

> "The presence of quantities of flint and obsidian chips in fill overlying tomb construction is a

> distinctive trait of Maya burial practices in the Peten and numerous instances [have] been found at Tikal in association with burials."
>
> (Trik 1963:3)

The most common material used in this deposit is chert flakes. In those cases where obsidian has been reported, it was deposited in alternate layers with chert (Coe 1988; Moholy-Nagy 1994, 1997), and in the case of Ka'Kabish, in alternate layers with carbon (personal observation).

It should be noted that not all burials at a site where this practice is evident will include these deposits. Reasons for the use of these materials in the filling of a tomb shaft have been discussed by both Coe (1988) and Moholy-Nagy (1997), who presented rather disparate interpretations of this event. Coe interprets the inclusion of chert and obsidian material in the sealing of a tomb to reflect the Maya belief in the underworld, the chert and obsidian pieces representing the House of Knives, one of the places endured by the Hero Twins during their initial visit to Xibalba (1988). Moholy-Nagy (1997) takes a more pragmatic position, viewing the inclusion of this material as a fortuitously utilised occasion to dispose of production debitage. While I am not willing to speculate on the ideological nature of a behaviour pattern as little recognised as this, I am unable to support Moholy-Nagy's argument that the use of the materials is the result of an opportunity to dispose of debitage (1997). While the chert pieces recovered *appear* to fit the general criterion of debitage (large primary and secondary flakes), the obsidian material noted at Ka'Kabish does not. Here the pieces conform with standard prismatic blades, albeit not particularly skilfully made ones, but nonetheless fully useful. The chert flakes used in the tomb shafts at Blue Creek and La Milpa are morphologically homogeneous, and therefore probably specially produced for filling the shaft (Guderjan 1998: pers. comm.). In addition, if, as Moholy-Nagy (1997) proposes, the purpose of including this material was merely to fill the excavated shaft, then the material initially removed from the area should be sufficient for the purpose, and the addition of other 'disposal material' would necessitate the removal of other ballast material that would then have to be disposed of elsewhere. The obsidian and chert are arranged in alternating layers, interspersed with ballast and other materials, indicating that the material was deposited with some forethought and not dumped haphazardly into the open shaft.

The care clearly taken in the placement of these layers, coupled with the functional value of the obsidian pieces included in the deposits, make it very doubtful that the deposits were the result of community waste management. Becker suggests that the interment of rulers in buildings conveyed a sense of identity, or sacred power, to the structure (Becker 1992). This concurs with McAnany's (1995) belief in the importance of "the continued physical presence of buried ancestors" and the "physical use and reuse of space" as a means of expressing genealogies and legitimising authority (McAnany 1995:8). As these deposits only appear in certain tombs constructed for elite individuals in public, ritual architecture, and consume quantities of both local and non-local resources, it is more likely that the inclusion of these materials was a deliberate part of the burial ritual. Consequently, while one might not be able to determine the *exact* nature of the deposits included in the filling of the tomb shafts, one can infer that it was a part of the burial ritual and is indicative of the status of the individual.

It is the opinion of this author that the material deposited in the tomb shafts indicated an interment ritual not clearly identified in the archaeological literature, and one warranting more study. However, for the purpose of this study, acknowledgement of its existence and potential ritual aspects is deemed sufficient.

**Ploughzone/Surface (PZ/S)**

The last 50 years has seen a resettling of the Blue Creek area by Mennonite populations. Consequently, much of the area to the west of the site, and most of the area to the east, off of the escarpment, has been developed for agricultural and ranching purposes. Due to the severe disturbance resulting from ploughing and discing activities, numerous artefacts were found in unclear context. As their original positions are indeterminable, it was decided to identify these artefacts on the basis of their recovery context.

With few exceptions, these artefacts are exclusively non-elite items. The majority of objects found in the developed area to the west are associated with the remains of plaster floors, while the area to the east contains the remnants of both plaster floors and low, single course platforms. Both of these surfaces would have been the foundation for simple pole and thatch structures.

While the nature of the superstructure used at the two types of residences may be similar, the amount of labour invested in the overall construction of the domicile is clearly different. Both of the residence types are believed to indicate non-elite status; however, they were deemed to represent two different strata of the community. Consequently, the social category of *non-elite* was separated into two levels, Non-Elite 1 and Non-Elite 2. The first of these two categories refers to those residences possessing a platform, while the second is used for those dwellings indicated by artefact scatters or plaster floor remains.

**Special Deposits (SD)**

This term is more in the nature of a catch-all than a specific event, and refers to deposits recovered at the site of Blue Creek that are currently beyond clear identification. Four of these deposits are known from various locations in the site core and settlement area. These deposits consists of large quantities of broken objects, usually distributed across the lower portion of a structure, or, as is the case with the Structure 13 Courtyard, strewn on the platform adjacent to a structure. As the deposits occur in both public and private locales, material consumed by these events are divided socially by the location of the deposit. Items found in private locales are deemed to be the possession of the individuals in residence, or in social association with those individuals. Objects found in public areas are problematic, as it is impossible to determine the social status of the objects' donor.

If one accepts the premise that the public structures are symbols of a community's identity, then it would follow that the events oriented around the role or function of the structure, or the termination of the use of the structure, would also be community events. Such events would entail participation, if not by every member of the community, then by at least a representative of each household or lineage group. It is likely that each participant made an offering of some type, with the most common appearing to have been ceramic vessels. Whether the vessels were the intended offerings or simply the receptacles for offerings of perishable materials is uncertain. What is unusual is the variation of the items included in these deposits. The deposit from Structure 3 at Blue Creek revealed items ranging from small objects, such as obsidian blades and jade beads, to larger polychrome and the occasional unslipped vessels. The disparity in the items, most notable in the quality of the vessels, supports the hypothesis that the materials came from a wide cross-section of the social spectrum, with everyone contributing the best item they were able.

Garber (1989) refers to these deposits as *'termination rituals'*. However, this term implies an understanding of the behaviour formulating the event, and highlights only one possible type of event. As the excavations of these deposits and the subsequent laboratory work have not yet been completed, it would be precipitous to discuss this material in terms of its ideological role. Consequently, the term *Special Deposit* is used in this report as this term refers more to the excavation context of the material than its possible social or ritual function.

**Workshops (WS)**

The reports regarding site within this study that use this term are Tikal and Colha (Driess and Brown 1989; Moholy-Nagy 1990). While the term is usually applied to a place of "formal tool manufacture" (Hester and Shafer 1984:165), this title is a slight misnomer in that the areas identified are not necessarily the areas of production, but rather the deposits of post-production debitage. At the site of Colha, "the workshop mounds consist solely of waste flakes, tools broken during manufacture and occasionally potsherds" (Hester *et al.* 1983:50). During the Late Classic Period at Colha, some of these mounds are "off the edges of the platforms where the chipping waste was apparently dumped" (Hester *et al.* 1983:50). While one might debate the exact definition of the term workshop (area of production versus specialised dump), areas so named are undeniably evidence of large-scale tool production. Only a few sites have reported specialised obsidian workshops (Kidder *et al.* 1978; Neivens and Libby 1976; Ford *et al.* 1997; Moholy-Nagy 1990). As no such deposits have yet been reported from Blue Creek, it was deemed wisest to let this term stand as it is used in other reports.

**Other (OTH)**

This term is used as a catch-all to allow items that have no clear provenience, either archaeological or functional, to be included in this analysis. These include material from disturbed contexts, backfill piles, or otherwise unknown contexts. The rationale for including such material in this analysis is that while an exact location is uncertain, these pieces still account for part of the total obsidian imported into a site and are therefore an important part of the archaeological record.

### FUNCTIONAL CONTEXT

In this model the material is grouped into larger, base units defined on the perceived function or nature of the excavation contexts. Five contexts were identified for this purpose. These contexts include: 1) ritual, 2) waste, 3) domestic, 4) burial, and 5) indeterminate.

Caches, Special Deposits and Tomb Shafts were considered 'ritual' activities and as such material from these contexts were grouped together under the category Ritual. Waste contexts were considered to be Construction fill, Collapse, Humus, Humus/Mixed matrices, Midden and Workshop deposits. Domestic contexts were identified as areas found either directly upon Floor Surfaces or within the Floor Fill.

Mortuary Goods were considered to be a separate category from Ritual or Domestic contexts by virtue of their indeterminate nature. While some items, such as *lak* plates with 'kill holes' may safely be assumed to be the possession of the interred individual other items are more problematic. It is virtually impossible to identify whether small objects such as blades or beads, as were owned by the individual or presented to him as a grave offering by another individual. All that can be said with reasonable certainty is that the individual in question was of sufficient standing in the community to warrant the removal of a certain amount of material from the systemic context. As such the material contained in burials is considered to be of separate function from other deposits.

The final context utilised in this model is for material from problematic or indeterminate contexts. Such contexts included Plough-zone/Surface matrices as well as materials found in Other or Unknown deposits.

## SOCIAL CONTEXT

This section attempts to define the possible social context in which the artefacts were found. As this type of context can be highly controversial, many reports prefer to omit references to the possible related social status of a deposit, choosing instead to concentrate on the archaeological, or excavation, context. However, such restrictions have ensuing impacts on the interpretation of the data. As it is likely that an individual's access to non-local, or luxury, materials is a direct reflection of that individual's social status, analysis of the social context, and the implied social status of the artefacts' possessors, are vital for understanding intra-site distribution patterns.

As many of the previously published reports did not focus on the context of the artefacts recovered, problems arose in identifying the nature of the different deposits. In order to overcome this limitation, a very general model for social divisions was created, consisting of nine categories. These categories are as follows: Elite 1, Elite 2, Non-Elite 1, Non-Elite 2, Non-Elite 3, Ritual/Ceremonial 1, Ritual/Ceremonial 2, Workshops, and Indeterminate. A full description of each of these categories is included below. However, it should be stated that while these categories may not reflect all the nuances found in a society as complex as the Maya undoubtedly were, they are, in the opinion of this author, the most applicable to the archaeological evidence available for this analysis.

### Elite 1 (E1)

Structures identified as belonging to this group were those residential structures in the core area of the site dating to the Early or Late Classic Periods. These structures have previously been identified as 'palaces' or 'range structures' (Bullard 1960; Loten and Pendergast 1984). It is the latter terminology that is used here. The range structures were constructed on low substructures or basal platforms, and possessed stone walls and either masonry or thatch roofs. They can occur singly or, more commonly, arranged around a small residential courtyard. The most notable characteristic that distinguishes them from the Elite 2 category is their location within the core area of the site.

Artefacts that were deemed to reflect the nature of the occupants, and therefore to be indicative of distribution patterns, were those found in primary deposits. These include items located on residential floors, included in caches interred within the residential structure or courtyard, and special deposits located within, or in front of, structures enclosed in the courtyard. The reason for including materials of an otherwise ritual nature in this category is that these deposits appear to have been of a private nature, and probably consisted of materials owned by the residents of the structure or courtyard, and not derived from public reserve.

Materials not included in this category were those items discovered in humus, collapse, or construction fill, as these may be the result of secondary deposits of recycled refuse or construction material. A more complete description of this practice is included above in the discussion of construction fill context.

### Elite 2 (E2)

Like the structures in the previous category, these structures fit the definition given for range structure. Like those found in the core, these structures were constructed on low substructures or basal platforms, and possessed stone walls and either masonry or thatch roofs. However, unlike those structures found in the core, these structures are found almost exclusively in groups of two or more arranged around small courtyards. The primary difference between the Elite 1 and Elite 2 categories is that those belonging in the second category are located outside the site core, away from monumental public architecture. In the case of Blue Creek, where personal observations of the area were possible, these courtyard groups were found to be located exclusively on limestone protrusions, jutting out from the fields around the site. The location of these courtyards on top of these promontories sets them apart from other, non-masonry, residences identified in the fields as much as their type of architecture. It is clear from the amount of labour invested in the structures that the individuals occupying these structures had greater access to labour resources than the occupants of the residences in the surrounding fields, where structures consisted of plaster floors and pole and thatch houses. Artefacts and deposit types included in this heading are the same as those outlined above under the heading of Elite 1.

### Non-Elite 1 (NE1)

Structures identified as NE1 possessed a raised stone platform upon which a semi-permanent structure of pole and thatched roofs would have been constructed. These residences are believed to represent a status of individuals who, while ranking higher than people who lived in less labour-intensive dwellings, did not possess the resources to construct the permanent dwellings that distinguish the two groups noted above. This category differs from that of NE2 in that structures that fall into this category are found within the core area.

It should also be noted that during the Formative Period, the actual status of the occupants of these platforms may actually equate to that of the Classic Period E1 residents. However, as these definitions are constrained by limited interpretative material and rely primarily on architectural evidence for status ascription it is necessary to identify them in the database as NE1. Artefacts and deposit types included in this heading are the same as those outlined above under the heading of Elite 1.

**Non-Elite 2 (NE2)**
These structures are identical in physical construction to those of their NE1 counterparts. The differentiating criterion for these categories is the location of the platform. NE2 structures are located away from the core, or other elevated areas, in lower, more agricultural, zones. As with the preceding NE1 category, consideration of the time period must be made when examining material within this group.

**Non-Elite 3 (NE3)**
This category of residences are often the most difficult to identify, as, unlike those of elite or even higher-status non-elite individuals, these houses were built with a minimum of labour investment, possessing no stone platforms or other masonry that would indicate occupation (Tourtellout 1988; Yaeger 1992). The only surviving evidence of the existence of these structures are the remnants of plaster floors and it is only through clearing and ploughing of the terrain that these residences become visible. Blue Creek Ruin is fortunate in that the surrounding area has been cleared on two sides, allowing for detailed maps and investigation of these areas (Baker 1996; Clagett n.d.; Lichtenstein 1997). The ploughed fields surrounding the site have revealed a number of white ovals, characteristic remains of this type of residence, and a comprehensive survey of the area is currently underway (Baker 1996; Lichtenstein 1997). Artefacts and deposit types included in this heading are the same as those outlined above under the heading of Elite 1.

**Ritual/Ceremonial 1 (R/C1)**
This category is designed to identify artefacts that were used in public rituals such as caches. It should be noted that one of the qualifiers of this category is that the material was used for the public, or community, benefit. Caches, as discussed above, are a ubiquitous event, found at virtually all Maya sites, and in almost all social stratum, differing visibly only in terms of type and quality of artefacts present. For a cache to be considered in this category it must be present in a public structure or place, an example being caches placed in plaza floors in front of structures. These caches differ from those placed in residential areas as the material used in the latter are probably composed of items available to, or directly possessed by the residents; caches in public contexts may be derived from a larger pool of public, disposable wealth, material held aside by the rulers for use for the good of the community (offerings to the gods, etc.), or may consist of objects provided by rulers or other elites are part of their social and ritual obligation.

**Ritual/Ceremonial 2 (R/C2)**
Public and ceremonial deposits of a more problematic nature than those outlined above were separated into their own category. In these cases, the nature of the deposit is unclear, as is the social origin of the material. Such cases include deposits in tomb shafts, material used as a layer, or bed, across the floor of a tomb, and artefacts distributed across the face or base of a public structure as described above under the heading *Special Deposits*. These materials may have come from public or elite stores -- collections of material reserved for use in public displays -- or they may result from individual donations by members of the community and therefore originate in a variety of different social stratum. The wide variety in quality of ceramic vessels discovered in the 'Special Deposit' in front of Structure 3 at Blue Creek (everything from polychromes to unslipped-unstriated vessels) would tend to support the latter hypothesis, that members from different social stratum were active participants in certain public rituals.

**Indeterminate (IND)**
As material included in the construction fill may be the result of recycled midden deposits, items found in construction contexts were deemed to be of indeterminate social context. Additional contexts deemed indeterminate were collapsed material, humus layer, and items classified as belonging to 'other' discussed above under excavation contexts.

**Workshop (WS)**
Controversy over the social implications of workshops, whether they were independent, household-based crafts or professions supported by elite patronage -- coupled with the difficulty in differentiating between primary deposits and secondary, specialised dumps, compelled me to leave this category without a social context.

## OBSIDIAN SOURCE IDENTIFICATION

**Purpose of Identification**
The third criterion included in this analysis is information regarding the original source of the obsidian used in the production of the artefacts. Analyses of the geological source of obsidian artefacts have been an accepted part of Maya lithic studies since the 1960s. The primary purpose of identifying the original source of raw material is to aid in the identification of possible trade networks (Hammond 1972, 1976, 1978; Sidrys 1977). Obsidian artefacts provide a perfect material with which to conduct source analysis for several reasons (*cf.* Braswell 1996:65-

67). The first of these is that the deposits utilised by the Maya have been identified, unlike jadeite deposits which are still largely unknown, or whose locations for various reasons have not been disclosed. Each obsidian deposit possesses a unique 'element fingerprint', making chemical identification an easy, and accurate process (Braswell 1996; Glascock 1998; Glascock *et al.* 1994).

Furthermore, artefacts composed of obsidian are ubiquitous throughout the Maya region, making them ideal for inter-site comparisons of source utilisation (Clark and Lee 1982; Driess and Brown 1989; Sidrys 1976; Sidrys and Kimberlin 1979; Zeitlan 1978). However, the majority of obsidian reports focus on the material from an individual site, or survey area, preferring to provide a temporal distribution of the material at that site rather than comparative discussions (Hammond 1976; Hammond *et al.* 1984; Healy *et al.* 1984; Guderjan *et al.* 1989; McKillop 1989; Mitchum 1994); Moholy-Nagy 1975, 1976; Moholy-Nagy and Nelson 1987, 1990; Rice 1984; Rice *et al.* 1985). Examination of the results of these studies have revealed that utilisation of obsidian sources was not constant for each source, but rather varied throughout the different time periods. Comparative analysis indicates an overall pattern where San Martin Jilotepeque dominated the Late Formative Period, giving way to El Chayal material in the Early and Late Classic, and yielding to Ixtepeque in the Terminal and Post Classic Periods (Dreiss and Brown 1989; Table 3.1 this volume). While this temporal pattern of source utilisation is widely accepted (Guderjan *et al.* 1989; Mitchum 1994; Moholy-Nagy and Nelson 1987, 1990; Rice *et al.* 1985), explanations for the dominance of one source over another are rare. When explanations are given they are characterised by hypotheses that discuss political machinations revolving around source competition, modes of production, and control of trade routes. It has been suggested that the shift from San Martin Jilotepeque to El Chayal material at the end of the Formative Period was the result of the increased political authority of the rulers at Kaminaljuyu, who "shut down" the production area around San Martin Jilotepeque to promote their resource base at El Chayal (Arnaud 1990; Brown 1977; Michel 1976; Nelson 1985; Sanders 1977; Sidrys and Kimberlin 1979). This theory has recently been called into question by new evidence suggesting that San Martin Jilotepeque was an active production area from the Formative Period through the Early and Late Classic Periods (Braswell 1996). Hammond (1972) suggested that the shift in emphasis was due to El Chayal traders utilising coastal and river routes, granting them wider and faster access to sites, while San Martin Jilotepeque merchants relied upon interior routes based on a combination of overland passes and the Chixoy-Passion River.

One potential aspect of the shift in source utilisation which has not been investigated is that of possible intra-site factors. No analysis could be found in the current literature correlating source utilisation with intra-site distribution. Crabtree (1968) noted the difficulties in blade production resulting from the presence of inclusions. A cursory examination of material from the different sources clearly demonstrates different concentrations of inclusions. It is possible that these inclusions, which may affect the quality of the blades produced, also affected the material chosen for various purposes, or accessible to different segments of a community. It is with this analysis in mind that materials from various deposits at Blue Creek and Ka'Kabish were subjected to neutron activation analysis at the Missouri University Research Reactor (MURR), the results of which were combined with information from other sites in Northern Belize and the North-eastern Peten to provide a larger regional basis upon which intra-site distribution of artefacts manufactured from distinct obsidian sources could be examined.

**Introduction to Obsidian Source Analysis**

Obsidian is a naturally occurring glass created by volcanic action. This action allows for two very disparate, and rather contradictory processes to occur: first, volatile components are able to escape, a process which involves the rock cooling slowly, and second, the formation of crystals is prevented, an event which occurs when the rock cools rapidly (Glascock 1994). When a balance between these actions occurs (i.e., when there is a rate of cooling slow enough to allow for the emission of the volatile components but rapid enough to preclude the formation of a crystalline structure, the result is obsidian. The absence of a crystalline structure offers stone craftsmen a material without pre-determined planes of cleavage that fractures under less force than chert. This absence of crystalline formations also gives the obsidian an edge so sharp it is "cleaved to the last particle of matter" (Crabtree 1968:471).

Obsidian sources in Mesoamerica may be divided into two areas: Mexico and the Guatemalan Highlands. The Mexican sources stretch from the Pacific Coast of Jalisco and Nayarit east through northern Michoacan into north-central Veracruz. The Guatemalan sources start in Chiapas along the Mexican-Guatemalan border and extend south-east through the Sierra Madres and into the south-west corner of Honduras (figure 2.3). While these areas contain numerous obsidian deposits, only a few were utilised by Mesoamerican populations (Sidrys *et al.* 1972).

Discussions of obsidian sources are often confusing, due to the lack of standardised terms. Sidrys, Andresen and Marcucci (1972:1) define a *source area* as "a large area containing several outcrops of obsidian . . . and which may or may not have similar chemical characterisations", while an *outcrop* is identified as "a single geological obsidian deposit with well-defined boundaries". Identification of artefacts to their original source has implications for understanding inter-site relations, resource procurement, and distribution networks. Consequently, many tests have been derived to accomplish this task. They include fission track analysis (Durannie *et al.* 1971), measurement of artefact density (Reeves and Armitage 1973), thermoluminescence (Huntley and Bailey 1978), Mössbauer spectroscopy (Longworth and Warren 1979), and measurements of natural radioactivity (Leach *et al.* 1983). However, none of these techniques has proven as reliable, nor become so standardised, as analysis of the chemical composition of sources and artefacts. Obsidian is composed of 70 per cent to 75 per cent silica ($SiO_2$), 10 per cent to 15 per cent Aluminium oxide ($Al_2O_3$), 3 per cent to 5 per cent Sodium Oxide ($Na_2O$), 2 per cent to 5 per cent Potassium Oxide ($K_2O$), 1 per cent to 5 per cent Iron Oxide ($Fe_2O_2$ and FeO), and small amounts, 0.1 per cent to 0.5 per cent of water ($H_2O$).

When this material is expelled in molten form into the surrounding region, it attracts and retains traces of various mineral elements from the surrounding rocks. As these mineral elements are collected, each obsidian source acquires a distinctive composition of elements. These elements form the basis for the chemical compositional analysis. Although variations are greatest between different obsidian sources, lesser variations may occur between different outcrops within the same source, allowing for the identification of material to specific areas or quarries (Braswell 1992, 1996:124-126; Glascock *et al.* 1988). Two types of chemical analysis prevail in obsidian sourcing studies. These are X-ray Fluorescence Spectrometry (XRF) and Neutron Activation Analysis (NAA). Virtually all Mesoamerican obsidian analyses have employed one of these two techniques, with the majority of analyses conducted in the last decade using the NAA procedure. A third analysis technique, Proton Induced X-Ray Emission (PIXE) was also used, although less frequently.

Another option, visual sourcing, has been battered about with mixed success and support since the early part of the twentieth century (Fuller 1927; Cressey 1974; Jackson and Love 1991; McKillop *et al.* 1988; Moholy-Nagy and Nelson 1990; Snow 1972 in Cressey 1974). This type of analysis relies on physical examination of the artefacts, with criteria based on visual characteristics, such as colour and inclusions. As this is not a science-based procedure, it is more open to criticism, and has acquired many sceptics (Glascock 1994; Moholy-Nagy and Nelson 1990). However, it is the opinion of this author that criticism of this technique is warranted only in those instances when its proponents have over-extended the capabilities of the procedure. Many of the researchers who have tried, and failed, to analyse their respective collections visually have been thwarted by their attempts to identify the *entire* test sample with this technique. Visual sourcing is a viable technique for artefacts exhibiting certain *distinctive* criteria, and is often used to identify material from Pachuca, know for its distinctive green colour (Dreiss 1989; Guderjan 1988; Holmes 1900; Rovner 1989; Spence 1996; Spence and Parsons 1967). However, visual sourcing should not be employed with the expectation of identifying every object in a collection. This factor should be kept fully in mind when using this technique, as will be discussed below.

As the material used in this report is derived from a variety of sites, it is to be expected that different testing techniques will have been employed. The material from Blue Creek was subjected to abbreviated NAA and visual sourcing techniques. Analyses from other sites employed NAA or XRF. These techniques will be discussed below.

**X-ray Fluorescence Spectrometry (XRF)**

X- ray fluorescence spectrometry is a technique for sourcing obsidian through irradiating samples with x-rays. This irradiation causes the displacement of atomic electrons from the inner, or lower, energy levels. Atoms from the outer, or higher, energy levels, seek to repopulate these vacancies. During the re-population process, energy is emitted in the form of x-rays. Dispersing the rays through a diffracting crystal permits them to be measured electronically. As each element emits a distinctive spectrum of x-rays, measurement of the types and intensities of the x-rays allows for the elements present to be identified (Glascock 1994; Jack and Heizer 1968; Weaver and Stross 1965). The elements used in sourcing obsidian vary depending on the laboratory performing the analysis, the year in which the study was conducted, and the opinion of the researchers as to which elements are most distinctive (Weaver and Stross 1965; Jack and Heizer 1968). However, some of the most common elements used are Barium (Ba), Iron (Fe), Manganese (Mn), Niobium (Nb), Potassium (K), Rubidium (Rb), Sodium (Na), Strontium (Sr), Titanium (Ti), Yttrium (Y), Zirconium (Zr) (Glascock 1994; Nelson *et al.* 1977).

X-Ray Fluorescence Spectrometry was popular during the 1970s, and many analyses of Mesoamerican obsidian were conducted using this process (Hammond 1972; Hester *et al.* 1971a, 1971b; Nelson *et al.* 1977; Stevenson *et al.* 1971). The advantages of this system over that of NAA are that it is less expensive and does not require the use of a nuclear reactor.

**Proton Induced X-Ray Emission Technique (PIXE)**

The proton induced x-ray emission (PIXE) technique is similar to XRF testing in that both use x-ray emissions as a means of measuring the proportions of various trace elements found in a material or artefact (Andrefsky 1998:43). However, where XRF excites the molecules using an X-ray beam, PIXE uses a proton beam (Walker 2000: personal communication). Both techniques use the resulting flourescence X-rays to measure the trace elements (CFANR; PIXE Analytical Laboratories; Walker 1999).

In PIXE testing, a proton beam is used to excite the electrons in the atom (CFANR; PIXE Analytical Laboratories; Walker 1999). Once excited, these electrons move away from the core, creating vacancies in the atom. The electrons do not remain separate but fall back into the spaces, a process that results in energies being created and subsequently emitted from the atoms in the form of flourescence X-rays. These flourescence X-rays are distinct, corresponding to the nature of the element which created them. The quantity of each energy is proportional to the quantity of the corresponding element present in the sample. Measurements of these energies create the necessary 'elemental fingerprint' required to identify obsidian artefacts to their original source (CFANR; PIXE Analytical Laboratories; Walker 1999).

Although both PIXE and XRF measure the emission of X-rays, XRF testing analyses the entire surface of an artefact, while PIXE, by virtue of the concentrated nature of the proton beam, can analysis a small area of the sample (Andrefsky 1998:43). This allows for different sections of the same artefact to be examined and compared for variations (Andrefsky 1998:43). PIXE is a non-destructive technique, unlike NAA which requires the specimen to be fractured into small pieces. However, it requires the specimens to be highly polished, a process which may result in damage to the surface of the artefact (Andrefsky 1998:43). This technique was used for testing of the Cuello obsidian collection (Hammond 1991b; Johnson 1976).

**Neutron Activation Analysis (NAA)**

As with XRF, NAA is a method by which obsidian artefacts may be traced to their original source. While NAA is more accurate than XRF, it is also a destructive technique. NAA is a qualitative and quantitative analysis that measures a number of major and minor trace elements in a sample.

Measurement of the trace elements is accomplished by irradiating the sample, then monitoring the gamma rays emitted during the decay periods. The exposure of the sample nuclei to neutrons produces excited compound nuclei. Such a nucleus will almost immediately decay, transforming into a stable, and often radioactive, isotope. The transformation is accomplished through the emission of *prompt gamma rays*. This radioactive nucleus continues the process of decay by emitting *delayed gamma rays*. The life of these rays will vary a mere fraction of a second to several years, depending upon the nature of the sample (Glascock *et al.* 1994).

As the gamma rays are emitted at two different intervals of the decay process, NAA may be divided into two different categories. The first is the measurement of the *prompt gamma-ray* emission during irradiation (PGNAA). The second is the measurement of the *delayed gamma-rays* emitted after the initial radioactive decay (DGNAA). PGNAA is used to measure those elements that produce only stable isotopes, possess weak decay emissions, or whose decay rate is too rapid to be measured by DGNAA (Glascock 1998).

DGNAA is the more common of the two techniques, and when NAA testing is mentioned, it may be assumed that this was the technique employed. This process has the advantage of time over the other method. Specifically, if the measurement of a desired radionuclide is impinged upon by the presence of another shorter-lived radionuclide, this process offers the option of waiting until the offending radionuclide has decayed before measuring the required radionuclide (Glascock 1997). Automated systems allow for the measuring of numerous elements simultaneously, a process that increases processing and accuracy. When instrumental procedures are employed in the analysis, the process is referred to as Instrumental Neutron Activation Analysis (INAA) (Glascock 1998; Glascock *et al.* 1994).

Material from Blue Creek was sent to the Missouri University Research Reactor (MURR). Under the direction of Michael Glascock, MURR has been engaged in a long term project to document, through NAA, all known sources of Mesoamerican obsidian (Glascock 1994; Glascock *et al.* 1994). The MURR database includes chemical fingerprints for more than 800 source specimens from roughly 40 sources throughout Mesoamerica. Full analysis of obsidian samples includes counts for 26 elements. These elements are as follows: Antimony (Sb), Barium (Ba), Cerium (Ce), Cesium (Cs), Chlorine (Cl), Cobalt (Co), Dysprosium (Dy), Europium (Eu), Hafnium (Hf), Iron (Fe), Lanthanum (La), Lutetium (Lu), Manganese (Mn), Neodymium (Nd), Potassium (K), Rubidium (Rb), Samarium (Sm), Scandium (Sc), Sodium (Na), Strontium (Sr), Tantalum (Ta), Terbium (Tb), Thorium (Th), Uranium (U), Ytterbium (Yb), Zinc (Zn), Zirconium (Zr) (Glascock 1994; Glascock *et al.* 1994). By tracking the appearance of various traits in different samples, researchers at MURR were able to isolate five elements that may be used to differentiate between Mesoamerican sources. These five elements are, in decreasing order of importance, Manganese (Mn), Sodium (Na), Barium (Ba), Chlorine (Cl), and Dysprosium (Dy) (Glascock 1994; Glascock *et al.* 1994). This process of using only five elements is referred to as *abbreviated neutron activation analysis*. This procedure identifies the source of artefacts with nearly 100% certainty, and may be used successfully to identify 90% of Mesoamerican obsidian artefacts (Glascock *et al.* 1994). For objects that cannot be assigned to a source through this technique, a full NAA test is still possible.

INAA was used to perform the abbreviated NAA tests on the Blue Creek material. Test samples were prepared by removing a small, 100 mg section from each artefact. These segments were weighed, fractured into chips, placed in high density poly-vials, and irradiated for five seconds, then allowed to decay for 25 minutes before the emissions were counted. The counting process lasted for 12 minutes and measured the five elements mentioned above (Glascock *et al.* 1994; Glascock 1998: personal communication). As the gamma ray emissions were recorded after the decay period, this is technically a DGNAA test. This process is standard for all NAA samples and would have been employed for samples from other sites analysed at MURR. The results of these tests are discussed in the following chapter under the relevant site headings.

**Visual Sourcing**

Attempts to visually identify obsidian to source is not a recent pursuit. Early attempts began in the late 1920s (Fuller 1927) and have continued (Clark *et al.* 1989; Clark and Lee 1984; Cressey 1974; Graham *et al* 1972, 1978; McKillop 1995) with sporadic support throughout the succeeding decades. With the advent of scientific tests such as those outlined above, visual sourcing techniques have been replaced in popularity and confidence, although the process still retains a few adherents (Clark *et al.* 1989; Cressey 1974; Jackson and Love 1991; McKillop 1995). Before the advent and widespread accessibility of chemical compositional analyses, visual characteristics were more frequently reported in discussions of artefacts (Graham *et al.* 1972, 1978; Jack and Heizer 1968; Kidder *et al.* 1978; Willey 1972). Both McKillop (1995) and Clark and Lee (1984, 1989) report great success with visual sourcing and have used it to identify complete collections.

Some articles used a combination of techniques, chemically sourcing a sample and then trying to determine visual characteristics shared by all pieces (Jack and Heizer 1968), or attempting to visually identify pieces and then testing them chemically to verify the results (Moholy-Nagy and Nelson 1987). The lack of success these tests exhibited is felt to be the result of the researchers trying to identify all the artefacts in their respective test samples. It is my opinion that the usefulness of visual sourcing lies not in its ability to *replace* chemical sourcing completely by allowing all the artefacts to be identified with this method, but in its ability to identify *some* of the pieces, thereby reducing the number of unknown pieces from which a sample must be sent for chemical testing.

Through the examination of artefacts identified by NAA, certain visual characteristics were able to be isolated and identified to source. Using these characteristics, the process was then reversed and pieces were visually identified then sent for chemical analysis to test and refine the criteria. With this process, the author has been able to create what are believed to be viable criteria for the three prominent obsidian sources utilised by the Maya. It should be noted that the description of visual characteristics presented here are those of the author, and do not necessarily compare with those used in other reports. They are, however, intended for possible future use by other researchers.

***Visual Sourcing Technique***

This procedure relies upon a physical examination of the visual properties of obsidian samples. The visual characteristics were divided into three primary categories: 1) colour, 2) inclusions, and 3) base matrix. Colour is a characteristic that has been used as a descriptive attribute in many obsidian reports (Holmes 1900; Spence and Parsons 1967; Tyler 1861). In some reports, it has been advocated as a means of creating source typologies (Cressey 1974). Many reports that otherwise do not advocate or use visual sourcing techniques do automatically assign any green obsidian as "Pachuca" in origin (Dreiss and Brown 1989; Guderjan 1988; Rovner 1989). This label is a slight misnomer as there are potentially three different outcrops of green obsidian within the Pachuca region (Cobean *et al.* 1967; Harlton 1969; Spence and Parson 1971, in Cressey 1974). In this study four colour varieties were identified: 1) green, 2) dark honey brown, 3) greyish black, and 4) black (Appendix I, figure 3).

Types of inclusions vary greatly, ranging from small rock particles scattered throughout the piece to misty 'clouds'. Rock, or discrete particle inclusions, are described in the following manner;

1) *large*: clearly visible with the naked eye, 0.7 mm and larger
2) *medium*: visible with the naked eye and a careful look, 0.3 to 0.7 mm
3) *small*: visible with the naked eye under close scrutiny, or clearly with a jeweller's loupe, 0.1 to 0.3 mm
4) tiny: visible with a 10x jeweller's loupe, less than 0.1 mm

The distribution of the inclusions also varies. The following groups of inclusions have been identified:

1) *band*: a layer that extends evenly through the piece. Often these layers are thin enough to almost disappear when looked at on edge.
2) *streak*: a collection of inclusions without clear parameters. These tend to fade or taper out around the edges of the clusters.
3) *trash*: a collection of particles of varying size and density that result in the piece becoming opaque. Trashy pieces can feel gritty to the touch. Tiny air bubbles, both in the interior as well as on the surface of these pieces, can contribute to the trashy appearance (Cox 1998: personal communication).

Other characteristics useful for purposes of visual identification include:

1) *clouds*: translucent areas of 'frozen smoke', or cobwebs: wispy forms of clouds. Clouds, or cobwebs, can be amorphous in shape, appearing randomly in the piece, or found sealed in bands.
2) *frozen ink:* a variation of clouds that appear "puffy" with billowing edges, giving the impression of injections of ink in water. Often, this type of inclusion can leave the piece solid black with only sections of clear glass showing
3) *milkiness:* these pieces are a translucent to opaque milky-grey, with no visible inclusions

While there are numerous other inclusions that can be identified as occurring in obsidian, the ones outlined above have been shown to be the most reliable visual indicators. It is important to note that not all pieces examined are visually identifiable. Due to overlap and variability in the characteristics, some pieces are too amorphous to be visually categorised. This paper is intended to present a method by which the sample can be culled, removing the obvious pieces and leaving only the indistinguishable pieces to be chemically tested.

The process for visually culling artefacts is outlined in Appendix I, figure 3. Briefly, the process involves first determining the base colour for the object. Once this has been done, certain pieces can be removed from further visual processing: green pieces can be assigned broadly to the Pachuca Region, while pieces that are completely black are considered suitable only for NAA testing. Pieces that are brown or greyish/black are evaluated on the basis of inclusions. Brown pieces with few large inclusions or streaks or small or tiny inclusions in a clear matrix are identifiable as Ixtepeque, while pieces with a milky opaque quality or high degree of 'trash' require NAA testing. Pieces that are grey/black in colour were found to contain three different types of inclusions: 'milkiness', 'trash', and 'frozen ink'. Of these only the milky pieces are considered securely assignable, and identified as El Chayal in origin.

In a test of 91 pieces sent to the MURR reactor, 54 pieces (59.3 per cent of the sample) were "culled" using this technique, while 37 pieces (40.7 per cent) were considered "non-identifiable by visual means". Fifty-two of these pieces were identified as 'definite' while two were considered 'questionable' assignments. Of the pieces examined, 52 were correctly assigned to source. This equates to an accuracy rate of 96 per cent. Only one mis-identified piece had been considered a 'definite' assignment. Using the method for visual sourcing outlined here, 57.1 per cent of the sample was successfully identified to source of origin, while visual sourcing was not deemed possible for 40.7 per cent pieces, and only 2.2 per cent of the pieces were mis-identified. Material analysed for this study came from the Blue Creek Ruin in north-western Belize. Chemical sourcing was done at MURR under the direction of Dr. Michael Glascock. Summaries of the pieces may be found in Appendix III.

### Obsidian Sources and Visually Recognisable Attributes

Due to the volcanic nature of the Sierra Madres, obsidian sources are numerous in the region. Obsidian sources in Central Mexico and in and around the Basin of Mexico (figure 2.4) include Pachuca, Ucareo, Zinapecuaro, Tulancingo, Paredon, Otumba, and Zaragoza (Charlton 1969; Ferriz 1985; Ericson and Kimberlin 1977; Glascock 1998; Healan 1993; Nelson *et al.* 1983; Spence and Parsons 1967). Sources from the Guatemalan Highlands (figure 2.3) include San Martin Jilotepeque, El Chayal, and Ixtepeque (Asaro *et al.* 1978; Coe and Flannery 1964; Michels 1975; Sheets 1975; Sidrys *et al.* 1976; Stross *et al.* 1977; Weaver and Stross 1965). Other sources of obsidian have been reported in the Guatemalan and Honduran regions, including Tajumulco, Jalapa, Sansare, San Bartolomé Milpas Altas, Laguna de Ayarza, Media Cuesta, La Esperanza and Guinope (Braswell 1996; Braswell and Glascock 1992; Sheets *et al.* 1990; Sidrys *et al.* 1976). However, many of these latter sites were either not worked by the Maya, or occur in such limited quantities throughout the Maya Lowlands that they are not considered to have been importance sources of supply.

While the majority of obsidian found at Maya centres comes from sources in the nearby Guatemalan Highlands, not all of the material recovered in the Maya Lowlands comes from this region. Pachuca, Zaragoza, and Tulancingo obsidian have been reported at a number of sites in the Maya region (Guderjan *et al.* 1989; Hanson 1990; Johnson 1976; Moholy-Nagy 1994; Moholy-Nagy and Nelson 1990; Nelson and Howard 1986; Spence 1996). The three most commonly reported sources at

Maya sites are Ixtepeque, El Chayal, and San Martin Jilotepeque (Appendix V). These are the sources that will be summarised here.

***Ixtepeque***

Ixtepeque obsidian begins appearing in small quantities at Maya sites in the Late Formative Period, and becomes the dominant source during the Post Classic Period. This source is located in south-eastern Guatemala, and spreads across the border into El Salvador (Coe and Flannery 1964; Michels 1975; Sidrys *et al.* 1976). The source area is approximately 300 sq. km, and comprises at least seven large outcrops. Numerous quarries and workshops have been reported for this source (Sidrys *et al.* 1976).

As the source is so large, it is only natural to expect variations in the characteristics. However, some of the material is clearly identifiable on the basis of colour and clarity. Obsidian that can be visually identified to this source is a rich, golden brown, with a high, glassy lustre and few if any inclusions. Inclusions in this source are rare and usually take one of two forms. The first type of inclusion is the occasional appearance of large particles, one or two in an otherwise transparent blade. The second type of inclusion found is bands or streaks of small particles that have a minimal effect on the transparent quality of the piece. The base matrix of the piece retains the same rich brown colour as those with few inclusions.

While the appearance of puffy, grey clouds have not been noted in this source, some pieces have been identified that possess a distinctive milky-opaque quality more common in the El Chayal source. However, even these pieces retain a brownish tinge that has not been apparent in the other sources examined. While these latter pieces retain the distinctive coloration, it is the opinion of this author that only the moredistinctive, clear, rich brown pieces should be assigned solely on the basis of the visual identification. Others should be considered part of the general collection requiring chemical testing. It is this procedure that was followed for the Blue Creek and Ka'Kabish material.

***El Chayal***

Material from this source first appears during the Formative Period, and later came to dominate the obsidian collections during the Classic Period (Table 3.1). The primary source area occupies 110 sq. km, and contains seven outcrops, two of which have been identified as possessing workshops. A series of non-outcrop related workshops also been identified in this area. Kaminaljuyu, the nearest major Maya site, has been identified as possessing several obsidian workshops, at which over 80 per cent of the material comes from the El Chayal outcrops (Michels 1976; Kidder *et al.* 1978:138-140).

El Chayal obsidian appears to have the most visual diversity. Pieces from this site range from gritty and filled with trash to clear with the occasional clouds and cobwebs. The most common characteristic, however, is a milky appearance. These pieces range in colour from dark to pale grey. While they are translucent, they have no visible inclusions; rather, the opacity is a result of the colour. Pieces from this source are generally very smooth, although the milkiness can affect the lustre, resulting in a dull or semi-gloss appearance.

Clouds and cobwebs also appear in this source, although the most distinctive of these inclusions are the "frozen ink" type. Inclusions of this type vary from small sections to encompassing the entire piece, with only a few small sections of clear glass showing. Usually, although certainly not always, the 'ink' is set into virtually clear glass. This particular variations may represent a distinct outcrop from the previously mentioned milky material.

It should be noted that sources reported elsewhere are known to include pieces that are composed of a base matrix that are "opaque black, [with] bituminous lustre" (Moholy-Nagy and Nelson 1987:16). These pieces are identified as coming from the Otumba, and possibly Ucareo, sources in Central Mexico (Moholy-Nagy and Nelson 1987). Consequently, pieces that are solidly or predominately black should be considered part of the general collection and considered for chemical testing.

***San Martin Jilotepeque***

This source is located along the Rio Pixcaya in Guatemala, and is sometimes referred to as Aldea Chatalun (Moholy-Nagy 1975). In 1976, Sidrys, Andresen, and Marcucci reported four outcrops, covering an area of roughly 60 sq. km for this source. This source dominated the Late Formative Period until it was replaced by the El Chayal source around the end of the Late Formative or beginning of the Early Classic Period (Table 3.1).

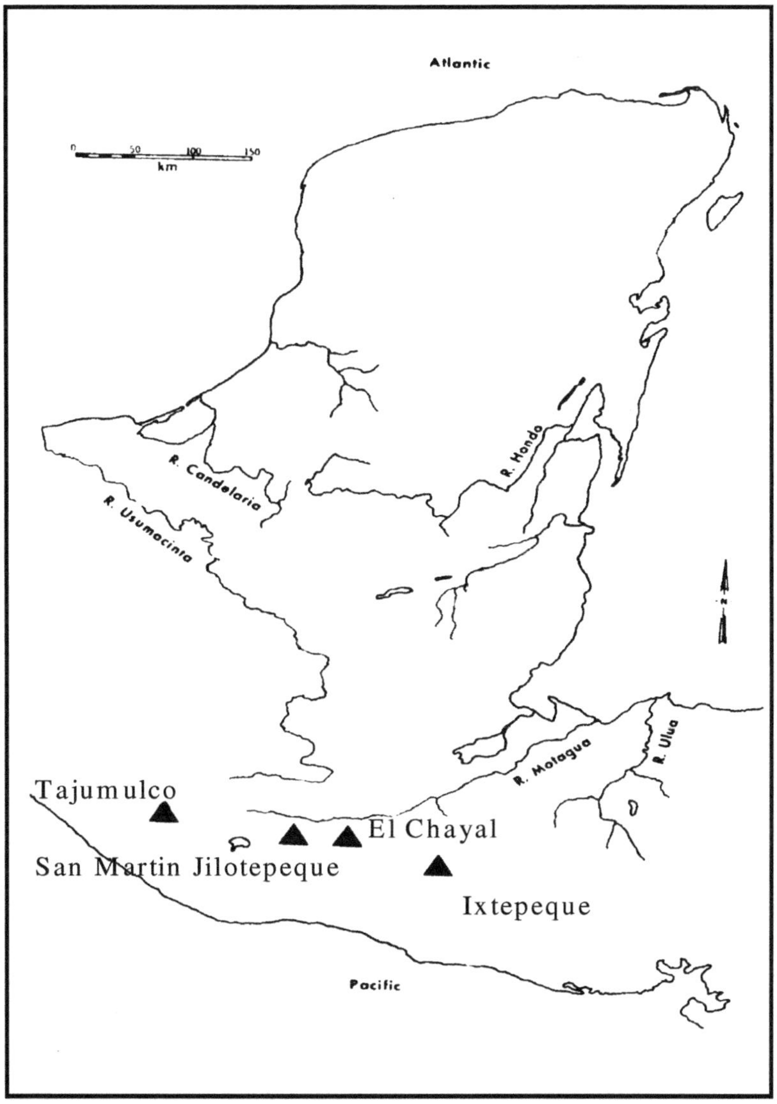


Map of Obsidian Source in the Guatemalan Highlands
Figure 2.2

*(adapted from Sidrys et al. 1976: figure 1)*

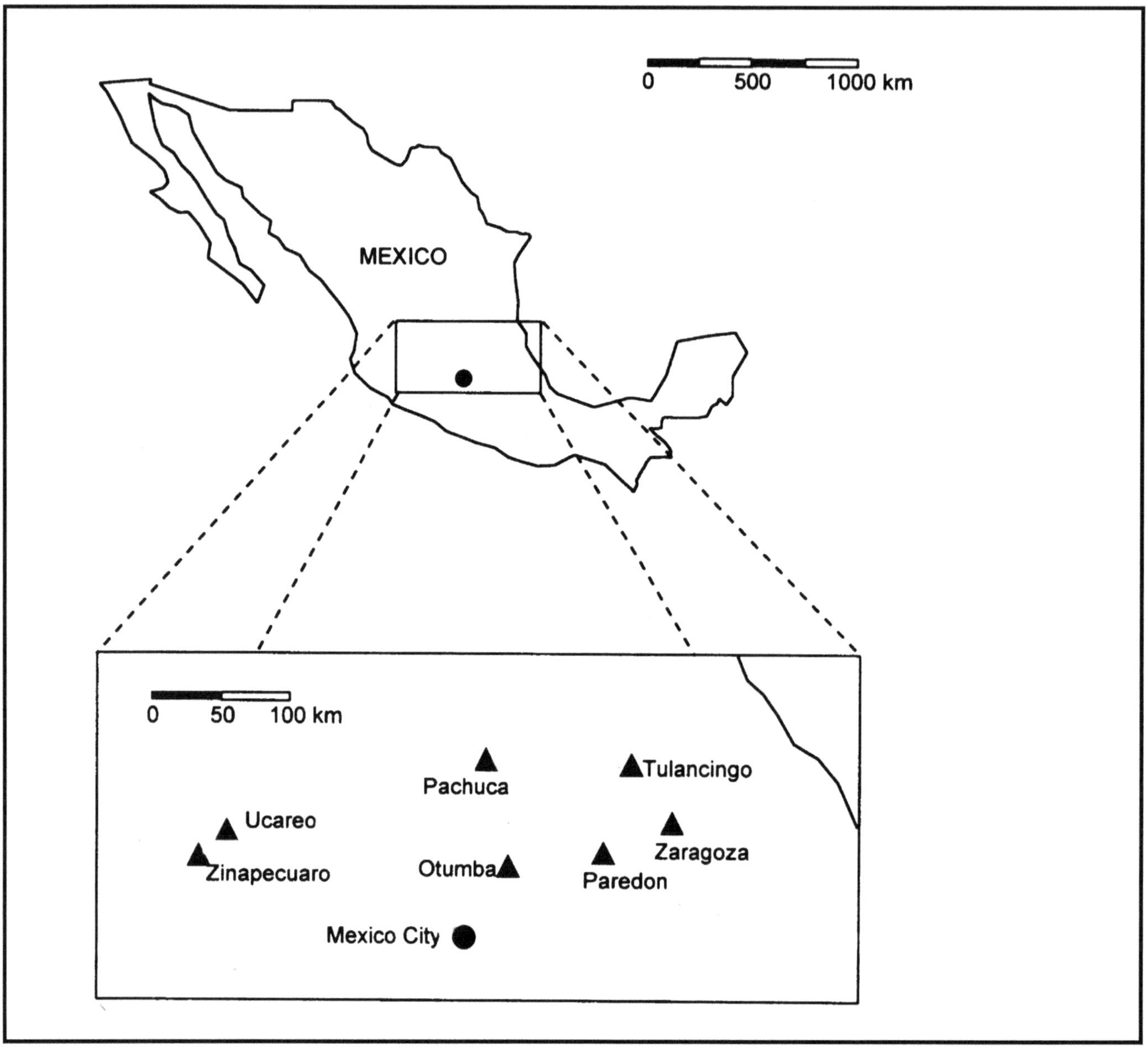


Map of Central Mexican Obsidian Sources
Figure 2. 3

*(adapted from Glascock 1998: figure 2.1)*

Pieces from this source can be described as trash-filled. They contain large quantities of particles of varying size and density, often mixed with grey clouds or cobwebs. Similar material has been chemically tested to the El Chayal source, making visual identification of these pieces by these criteria exceedingly difficult. The primary difference between the El Chayal and San Martin Jilotepeque material is that the latter appears with a great variety of inclusion in single pieces, and the lack of clarity is a result of the density of inclusions and not by colour or 'milkiness'. However, material that matches this descriptive pattern should be considered for chemical sourcing, and not assigned merely on the basis of visual attributes.

**Summary of Visual Sourcing Technique**

When analysing obsidian using visual techniques, the three most important aspects to consider, in order of importance, are colour, clarity, and inclusions. Ixtepeque material is the only source so far identified with a distinctive rich brown colour. El Chayal, while exhibiting a greater range of characteristics than Ixtepeque, can be most easily identified by its two primary types: 1) pieces with a milky matrix, regardless of the quantity or type of inclusion, and 2) pieces containing 'ink clouds' in a clear matrix. The San Martin Jilotepeque material is the most difficult to identify. While pieces with a high degree of trash set in a clear base matrix are common to this source, it is the opinion of this author that such pieces are best left in the general collection for chemical analysis.

While visual sourcing lacks the ability to accurately identify every piece of obsidian examined, it should not be discarded as an analytical tool. When used correctly, visual sourcing can be as useful tool in helping to increase the amount of obsidian sourced, and to decrease the size of the sample in need of chemical testing.

### Research Limitations

As this type of study is uncommon in archaeological obsidian reports, problems of compiling a suitable database were not unexpected. The most common problem encountered was the tendency for reports to provide limited information regarding the archaeological or systemic context of the artefacts. Those artefacts that were accompanied by contextual data revealed the tendency for material from only one type of deposit to be sampled. This limitation made it necessary to broaden the nature of the sample to include several sites from different regions. While this increases the number of artefacts in different contexts, ideal circumstances would have seen broader samples from more sites.

The problem of a limited number of sites that provided the necessary contextual information was further compounded by the difficulty in finding sites with a large enough database for comparison. The cost of obsidian sourcing results in a limited number of pieces being submitted for testing. While material from several sites was tested in small batches in different years, this only made the use of these articles more difficult, as authors sometimes conflated the numbers to include the previously tested material, and at other times they presented only the new data, leaving one to sort the data carefully and compare the numbers with previous papers before creating a usable database from the material. The numbers that are presented in the written reports are almost exclusively those for the number of artefacts analysed, with little or no mention made of the total number of pieces found, creating a lack of clarity in regards to the relative percentage of material tested and its respective implications for the obsidian consumption at the site as a whole. Although one might justifiably argue that the total quantities of various artefacts are data that may be found elsewhere (namely site monographs), these reports often post-date the publication of analysis articles by several years, making the inclusion of this information in articles that much more relevant.

Another challenge was the decision on a standardised measure by which to conduct the analysis. While the ideal measure for discussion of access and consumption patterns would be weight, this measure is rarely included in reports. As discussed above, the preferred means for presentation of obsidian data is per piece recovered. Since this is the format used in the reports from other sites, and in order to keep the data consistent, this will be the primary format of this report. However, in an effort to track all the possible nuances surrounding the distribution of obsidian material from Blue Creek and Ka'Kabish, data will also be presented in terms of weight and differences in quality. The application of these terms is discussed above under their appropriate headings.

## Section 2
## Sites Analysed

### Introduction

As obsidian is present at virtually all Maya sites at some point in time, studies such as these must have limits, lest their size make them more unwieldy than useful. Consequently, this study will be limited to Northern Belize and the Eastern Peten. El Mirador forms the western boundary of the research area, with the coast of Belize and off-shore Cayes forming the natural eastern limit. All sites with the one exception discussed below, fall between the Hondo and Belize Rivers, the northern and southern limits respectively. Although these parameters place a constraint on the size of the study, they still permit variability in site location. A total of twelve sites and two survey areas were selected for this study. The sites are as follows: Blue Creek, Becan, Cerros, Colha Cuello, El Mirador, Ka'Kabish, Kichpanha, Moho Caye, Nohmul, Northern River Lagoon and Tikal. The two survey areas included were the Tikal/Yaxha Corridor and the Central Peten Historical Ecology Project (CPHEP) referred to here as the Central Peten Lakes (Appendix I, figure 4).

As discussions of Maya trade patterns often involve predictions about trade routes and site-function as determinates in resource access and monopolisation it was felt that the sites chosen for this study should attempt to reflect the diverse settlement patterns discussed in these texts. Consequently the sites used in this analysis were divided into four broad categories: 1) coastal, 2) river oriented, 3) inland, and 4) Central Peten sites. While definitions for coastal and river oriented sites are relatively straight-forward, those of inland and Central Peten sites requires a slight clarification. Central Peten sites are those centres situated in the quadripartite junction of the Rios Passion, San Pedro Martin, Azul and the head-waters of the Belize River. This area forms the geographical centre of the Maya realm (Rathje 1972). Inland sites are those sites that are not located near adjacent water-ways, and whose access to such passages would have involved an overland trek or mediation through a second site.

The site of Becan falls outside the pre-defined northern limit of the Rio Hondo. However, this site fits the category of an inland site admirably. As there are only two other sites within the proscribed study area, both with limited collections, it was decided that the information available from Becan merited inclusion in this study. Furthermore, one might argue that as the Rio Hondo is the northern-most river providing access to the Yucatan interior, that Becan, by virtue of its proximity to the Rio Hondo watershed, could be considered a benefactor of trade goods shipped along this route. A full discussion of the sites, and their artefacts, is found below.

## Discussion of Sites Analysed

Although this report is designed to be a inter-site analysis, the basis for the research is derived from work conducted at the site of Blue Creek in North-Western Belize. With this consideration in mind the Blue Creek Ruin will be discussed first with the other sites included in their appropriate geographical classification below.

### Blue Creek

Located in North-Western Belize the Blue Creek Ruin is considered a medium sized Maya centre. Surveyed by Mary Neivens in 1976 (Neivens 1991), excavations at the site were not initiated until 1992 when the Maya Research Program under the direction of Dr. Thomas H. Guderjan obtained a permit for the site (Guderjan *et al.* 1993). Investigations quickly revealed that the size of the site far exceeded initial expectations. While the core area was mapped during the 1992 season, surveying and mapping of the surrounding area has been virtually continuous throughout the succeeding seasons (J. Baker 1997; R. Baker 1995, 1996; Guderjan 1995; Guderjan *et al.* 1993).

Situated atop the Rio Bravo Escarpment in North-Western Belize, the Blue Creek Ruin commands a spectacular view of the northern Belizean Coastal Plain. The confluences of the Rio Bravo and Blue Creek River are clearly visible as is the juncture where they join to form the Rio Hondo. These rivers were undoubtedly a prosperous trade route during the Classic Period, and were probably an important factor in the creation and settlement of the site. Also visible on the horizon are the sites of Ka'Kabish and Lamanai to the east and El Posito to the north-east. Inter-site relations between these sites is currently unclear. There is little question that these sites were aware of each other and interacted to a degree, however the exact nature of these interactions is open for future investigation.

The Blue Creek Ruin possesses two large plazas, aligned roughly north-south along the limestone ridge that surmounts the escarpment (Appendix I, figure 5). Plaza A, the southern-most of the two plazas, is ringed by six structures, identified numerically from one to six. Structure 1 is the largest structure at the site. This building appears to have been initiated in the early part of the Early Classic, and underwent several modifications through the succeeding centuries. Structures 2 and 3 are small pyramidal buildings that define the eastern perimeter of the plaza. Initial surface inspections suggested structures roughly similar in size and form. Structure 2 was severely looted, jeopardising the structural integrity. Consequently, investigation of the structure was constrained to defining the perimeter and its relation to Structure 3 to the south. At the south side of the plaza is Structure 4, a range structure that like Structure 1, appears to have been modified several times throughout the Classic period. Structure 4 is one of the oldest structures in this complex, with material dating to the latter facet of the Late Formative Period. Structures 5 and 6, located along the western edge of the plaza, are contemporaneous with Structure 4. Although both Structures 5 and 6 are range structures, Structure 5 is considerable larger than its northern neighbour, possessing a superstructure in excess of 40 metres.

Immediately to the north of Structure 1 is a low platform upon which Structures 7 and 8 form a small, open-ended ballcourt. To the east of these structures and between Plazas A and B is Structure 9. This is a large 11 metre tall structure situated upon a discrete 2 metre tall platform. Two ramps, one on the western edge the other on the northern connect this structure to Plaza A and the Structure 13 Courtyard respectively.

The Structure 13 Courtyard forms the southern boundary of Plaza B. This courtyard is a residential complex consisting of two range structures situated on substructural platforms, two rectangular masonry structures constructed directly on the courtyard surface and affixed to the larger range structures, a small, pyramid structure possibly used as an ancestor shrine and a small domestic structure consisting of a low masonry wall topped by poles and surmounted by a thatch roof. This last structure was appended to one of the rectangular masonry structures probably late in the use of the courtyard.

Plaza B, unlike Plaza A, is longer with less defined, open space and more residential compounds (Appendix I, figure 5). At the northern end of Plaza B is Structure 25, this is a steeply sided pyramid structure approximately 10 metres tall (Renaud 1999:47). Dubbed the *Temple of the Obsidian Warrior* by Neivens because of a large number of obsidian pieces discovered in conjunction with a looted tomb, this temple has suffered severely from illicit digging. All attempts to locate the obsidian collected from this structure by Neivens in 1976 have been unsuccessful. The Structure 19 courtyard complex is situated in the approximate centre of the plaza. This complex consists of a series of interconnected residential range structures arranged around two, small central courtyards. Several other structures are scattered around Plaza B, either singly or in pairs (Appendix I, figure 5).

The Structure 25 Courtyard is a residential compound consisting of three rectangular structures situated on a limestone outcrop to the east of the Structure 13 Courtyard. Consisting of three single room range structures this area is one of the oldest inhabited areas of the site. Midden deposits from the lowest levels of the platform indicate an occupation in the early part of the Late Formative Period.

The Blue Creek Ruin continues beyond the escarpment with settlements located both above and below the ridge (Appendix I, figure 5). To the north-west of the core are two residential complexes, identified as the Structure 37 and 41 Courtyards. Like the Structure 25 Courtyard these are arranged atop limestone outcrops. Of these two courtyards only the Structure 37 complex has been investigated. Below the escarpment, the occupation takes the form of low, residential platforms arranged in clusters. One large limestone tor located close to the Rio Hondo yielded midden occupation dating to the Late Formative although investigations failed to locate any structures. Both above and below the escarpment, simpler domiciles were discovered. These residences consisted of pole and thatch structures with plaster floors. The only visible remains of these structures are remnants of plaster floors, and these only in areas where the ground has been ploughed.

When work at the Blue Creek Ruin began, the site was thought to be a Late Classic outpost for a much larger site to the west. As a periphery town one would expect that only limited amounts of exotic or non-local goods would have been available to the populace. However, six years of excavation revealed many fallacies with this initial interpretation, and forced us to rethink many aspects of life at Blue Creek. We now know that the site was settled during the later part of the Middle Formative Period (Haines 1996, 1997; Haines and Suther 1997; Haines and Wilhelmy 1999), and possessed a notable population during the Late Formative Period (Clagett 1997, 1999). The Late Formative settlements included residential areas both above and below the escarpment (Clagett 1997, 1999; Haines 1996), along with incipient monumental architecture and ritual activity in Plaza A (Pastrana 1996; Weiss 1995, 1996).

During the Early Classic Period the civic centre was expanded with the majority of monumental architecture on Plaza A being constructed (Driver 1999). This period is marked by a ritual event that consumed a large amount of non-local material in the form of jade artefacts (Guderjan and Weiss 1995; Pastrana 1999; Weiss 1995). Along with revising our beliefs about the occupation of the site, we found that a re-examination of our interpretation of the economy was also warranted. As the site overlooks three water routes (Rios Bravo, Hondo and Azul), it is to be expected that the populace had access to non-local goods, transported along the rivers, in greater proportions than its size would warrant. An examination of artefacts classified as *Special Finds* (i.e. artefacts not falling into the general categories of ceramics, lithic debitage or general faunal remains), supports this theory. Approximately 4000 artefacts discovered at Blue Creek fall into this category, over half of which were made of non-local material. Obsidian artefacts comprise roughly one quarter of all *Special Finds*, recovered between 1992 and 1997, the analysis of which forms the basis of this report.

Obsidian artefacts were recovered from a variety of locations throughout the site. A total of 1127 artefacts were collected between 1992 and 1997. Material from Late Formative deposits made up the largest proportion of obsidian collected (Table 2.1). Almost 56 per cent, or 630 artefacts came from this period.

| Date | Total Pieces | Total Sourced | # NAA Identified | # Visually Identified |
|---|---|---|---|---|
| LF | 630 | 403 | 85 | 318 |
| EC | 268 | 157 | 37 | 120 |
| LC | 145 | 106 | 49 | 57 |
| ND | 84 | 60 | 37 | 23 |
| | 1127 | 726 | 208 | 518 |

Blue Creek Obsidian Artefacts By Time Period
Table 2.1

Of the 630 Formative artefacts 74.9 per cent (n=472) came from caches, with the majority of these, 469 (99.4 per cent), belonging to a single deposit discovered in front of Structure 4 (Weiss 1996).

| Date | Total | Total in Caches | Caches = x per cent of Total Sample |
|---|---|---|---|
| LF | 630 | 472 | 74.92 |
| EC | 268 | 243 | 90.67 |
| LC | 145 | 21 | 14.48 |
| ND | 84 | 1 | 1.19 |
| | 1127 | 737 | |

Quantities of Cached Obsidian Artefacts at Blue Creek
Table 2.2

The other 158 artefacts were dispersed through the following eight categories; construction fill (CF), floor fill (FF), floor surface (FS), midden (M), mortuary good (MG), other (OTH) and plough zone or surface (PZ/S). Midden material formed the second largest collection of material from this period, with 18.1 per cent (n=114) of the artefacts found belonging to this category.

As the majority of the material from the Formative Period came from the cache beneath Structure in Plaza A it follows that the largest social context category is Public Ritual/Ceremonial (R/C1) with 74.6 per cent of the collection (n=470). The second largest level of consumption was found in the Non-Elite 1 segment of the community. A total of 116 pieces (18.4 per cent) of the collection was recovered in Non-Elite 1 contexts (Table 3.46). Twenty pieces (3.2 per cent) was found in Non-Elite 2 contexts, and 22 pieces (3.5 per cent) were recovered from indeterminate contexts.

Deposits from the Early Classic Period span seven categories, several of them identical to those found in the Formative Period. These categories are cache (C), construction fill (CF), collapse (COL), floor fill (FF), floor surface (FS), mortuary good (MG), and other (OTH). As with the previous period, the majority of the artefacts (n=236, 88.1 per cent) were recovered from a single cache (Haines 1999). However, unlike the previous cache this deposit was centred in a residential courtyard complex adjacent to the core, the Structure 25 courtyard, and not a public structure (Haines 1999). Material from this period makes up roughly 24 per cent (n=268) of the total Blue Creek collection.

Although the majority of the Early Classic material was recovered in a cache, the location of the cache within a residential courtyard places this deposit in the 'Elite 1' social sphere. Consequently, the contexts in this period responsible for the largest consumption of obsidian are 'Elite', with 90.7 per cent (n=243) of the material belonging to this group. The second largest social context from this period are indeterminate contexts, although this category lags far behind that of Elite accounting for only 7.1 per cent (n=19) of the Formative collection.

Late Classic material formed the smallest datable collection with 145 artefacts (12.87 per cent) coming from this period. Artefacts in this period were recovered from the following contexts: cache (C), construction fill (CF), collapse (COL), floor fill (FF), floor surface (FS), humus (H), other (OTH), plough-zone/surface (PZ/S) and special deposit (SD). A few pieces (n=12) were recovered from mixed contexts, humus/construction fill, humus-collapse and humus-floor surface. Special deposits formed the largest of these categories with 74 artefacts (51 per cent) being recovered from collections of this nature. Only 21 artefacts, less than 15 per cent of the Late Classic collection were recovered from cache contexts. This represents a significant decline in material being utilised for caches from the Late Formative and Early Classic Periods (74.9 per cent and 88.1 per cent respectively).

Deposits from Elite 1 contexts amounted to 28.9 per cent (n=42), while those belonging to Elite 2 contexts accounted for 42.8 per cent (n=62) of the Late Classic collection. This continues the trend toward greater elite consumption, with a total of 104 artefacts (71.7 per cent) being recovered from elite contexts. Indeterminate contexts formed the next largest category with 26.9 per cent of the material (n=39).

Material from Blue Creek was sourced using two different methods, abbreviated neutron activation analysis (NAA) conducted by Michael Glascock at the MURR reactor, and visual sourcing conducted by the author. Both these techniques are outlined in the preceding Section 1. Of the 1127 artefacts in the Blue Creek collection 726 pieces, or 64.4 per cent of the total collection were sourced. Sourced material included 208 pieces identified using abbreviated NAA and 518 by visual analysis. Artefacts from dated deposits made up 91.3 per cent of the sourced collection (n=666) with 60 sourced pieces (8.3 per cent) being from undated deposits. The material from dated deposits included 403 artefacts from the Late Formative Period (55.5 per cen*t*), 157 pieces from the Early Classic (21.6 per cent) and 106 items from the Late Classic (14.6 per cent). Hereafter the undated material will be excluded from the analysis.

**Total Sourced Material**

The 403 Late Formative artefacts included 376 pieces manufactured from El Chayal obsidian (93.3 per cent of the Late Formative collection), 25 pieces from San Martin Jilotepeque (6.2 per cent), and 2 items produced from Ixtepeque material (0.5 per cent) (Table 2.3). While these percentages may seem unusually high in favour of the El Chayal source, the majority of those pieces (n=325) came from one cache. With this deposit factored out of the equation the percentages of obsidian from each source are 65.4 per cent from El Chayal, 32.1 per cent of San Martin Jilotepeque material and 2.6 per cent of Ixtepeque material. These ratios are in line with those found at Becan (Dreiss and Brown 1989; Rovner 1989), Edzna (Nelson *et al.* 1983), and El Mirador (Haines 1995, Hansen 1990). However, it is not unusually for sites in the Central Peten area to have larger percentages of San Martin Jilotepeque during this period (Moholy-Nagy 1975, 1984; Moholy-Nagy and Nelson 1987, 1990; Rice 1984; Rice *et al.* 1985).

Early Classic material from Blue Creek retains a high percentage representation of El Chayal material. Of the 157 pieces identified to this period, 152 (96.8 per cent) were determined to have been manufactured from El Chayal material, while 2 (1.3 per cent) were produced from San Martin Jilotepeque material and 3 (1.9 per cent) from Ixtepeque obsidian. As with the previous Late Formative Material the percentage for El Chayal material is skewed slightly due to the presence of a large, single-use obsidian deposit in this level. If the 139 El Chayal and one San Martin Jilotepeque objects belonging to this deposit are removed, the quantities of obsidian correspond with commonly documented patterns (Dreiss and Brown 1989; Moholy-Nagy 1975, 1984; Moholy-Nagy and Nelson 1987, 1990; Rice 1984; Rice *et al.* 1985; Rovner 1989). After excluding the cached objects, the quantities of El Chayal obsidian are reduced to 76.5 per cent, while the San Martin Jilotepeque and Ixtepeque contributions are 5.9 per cent and 17.6 per cent respectively.

Late Classic artefacts continue to demonstrate a predominance of El Chayal material with 98 of the 106 objects (92.5 per cent), being produced from this source. Objects manufactured from San Martin Jilotepeque material accounted for 3 items (2.8 per cent) and Ixtepeque for 4 objects (3.8 per cent). The Late Classic was the only period in which Mexican obsidian was discovered at Blue Creek. This was a single piece of Zaragoza obsidian which equated to 0.9 per cent of the Late Classic collection. Although this period, unlike the previous two, did not possess caches composed of high numbers of obsidian, it did include two Special Deposits that held a total of 74 pieces of El Chayal obsidian. With these deposits factored out the levels of obsidian drop to more typical ranges. El Chayal material then accounts for 75 per cent of the sample, San Martin Jilotepeque for 9.4 per cent, Ixtepeque for 12.5 per cent and the Mexican material for 3.1 per cent of the collection.

| Date | ELC | SMJ | IXT | MX | Total |
|---|---|---|---|---|---|
| LF | 376 | 25 | 2 | 0 | 403 |
| EC | 152 | 2 | 3 | 0 | 157 |
| LC | 98 | 3 | 4 | 1 | 106 |
| Total | 626 | 115 | 9 | 1 | 666 |

Total Material of Blue Creek Obsidian
by Source and Time Period

Table 2.3

**NAA Sourced Material**

Chemical identification of the 208 artefacts,approximately 18.5 per cent of the total sample, was conducted by Michael Glascock at the MURR Laboratory using neutron activation analysis (see Chapter 2, Section 1 this volume).

Material from all three time periods (Late Formative, Early Classic and Late Classic) were sent for analysis as were some pieces for which a temporal association could not be determined. The later pieces were sent as part of a previous study (Dreiss and Glascock 1995) and as such were chosen to suit a different research model. The non-dated pieces were excluded from the following analysis.

| Date | Total # Pieces | # NAA Tested | per cent of Period NAA Tested |
|---|---|---|---|
| LF | 630 | 85 | 13.49 |
| EC | 268 | 37 | 13.80 |
| LC | 145 | 49 | 33.79 |
| ND | 84 | 37 | 44.05 |
|  | 1127 | 208 | 18.5 |

Blue Creek Material Tested by NAA

Table 2.4

Of the 171 artefacts sent for analysis 85 (49.7 per cent) were from the Late Formative, 37 pieces (21.6 per cent) belonged to the Early Classic Period, while 49 artefacts (28.7 per cent) were identified as Late Classic.

| Date | ELC | SMJ | IXT | MX | Total |
|---|---|---|---|---|---|
| LF | 58 | 25 | 2 |  | 85 |
| EC | 33 | 2 | 2 |  | 37 |
| LC | 42 | 3 | 3 | 1 | 49 |
| Total | 133 | 30 | 7 | 1 | 171 |

Blue Creek NAA Tested Materia
by Source and Time Period

Table 2.5

Within the Late Formative, the majority of the 85 pieces were El Chayal in origin (68.2 per cent). Twenty-five pieces (29.4 per cent) were from San Martin Jilotepeque, and 2 (2.4 per cent) pieces were identified as coming from Ixtepeque. These ratios closely parallel those obtained with the inclusion of visually sourced material.

The Early Classic collection contained 37 pieces sourced by NAA. Of these pieces 33 (89.2 per cent) were identified as El Chayal in origin. Two pieces (5.4 per cent) were identified as coming from each the San Martin Jilotepeque and Ixtepeque sources.

The Late Classic sample was the first to contain material from Central Mexico. A single piece of Zaragoza obsidian (2.0 per cent) was found among collection of 49 pieces. As with the previous period, El Chayal material dominated the collection with 42 pieces (85.7 per cent). Again material from both San Martin Jilotepeque and Ixtepeque were evenly represented in this collection with each contributing three artefacts (6.1 per cent each).

Due to the financial constraints involved in NAA testing, it is possible to test only a portion of each deposit. While the ideal is a collection where equal percentages of each period are sampled, the Late Formative and Early Classic Periods, as discussed previously, contained large caches of obsidian. Deposits of this nature may skew interpretations of the material and present a disproportionate image of each period (Orton 1998: personal communication). When material from these deposits is included in the analysis, NAA material from these periods appears under-represented (Table 2.4). However, when these disproportionate deposits are factored out (Table 2.6), or included as single elements (Table 2.7), then a more balanced perspective of the period materials may be presented (Orton 1998: personal communication). In these cases the proportions of NAA sourced material for each period is 30 per cent or greater.

| Date | Total Pieces - Caches | Total NAA - Caches | per cent of Period NAA Tested (excluding Caches) |
|---|---|---|---|
| LF | 158 | 48 | 30.38 |
| EC | 25 | 14 | 56.00 |
| LC | 124 | 49 | 39.52 |

NAA Sourced Material
Excluding Cache Artefacts

Table 2.6

| Date | Total Pieces Caches = 1 ea. | Total NAA Caches = 1 ea. | per cent of Period NAA Tested (Caches = 1 ea.) |
|---|---|---|---|
| LF | 162 | 49 | 30.25 |
| EC | 29 | 17 | 58.62 |
| LC | 125 | 49 | 39.20 |

NAA Sourced Material
Including Caches as Single Items

Table 2.7

**Visually Sourced Material**
Five-hundred and eighteen artefacts were identified visually by the author using the process outlined in Section 1 of this chapter. Twenty-three objects had no temporal provenience and were discounted from further analysis. Late Formative objects accounted for the largest group of material identified, with 318 artefacts being identified to the El Chayal source (Table 2.8). This was the only source identified by this method in the Late Formative. Early Classic artefacts consisted of 119 pieces of El Chayal material, and one item identified as Ixtepeque. Late Classic material included object from these two sites, with 56 artefacts being identified as El Chayal and one piece as Ixtepeque in origin. Items with no temporal affiliation included 21 pieces of El Chayal obsidian and 2 items of Ixtepeque material.

| Date | ELC | SMJ | IXT | MX | Total |
|---|---|---|---|---|---|
| LF | 318 | 0 | 0 | 0 | 318 |
| EC | 119 | 0 | 1 | 0 | 120 |
| LC | 56 | 0 | 1 | 0 | 57 |
| Total | 493 | 0 | 2 | 0 | 495 |

Blue Creek Visually Sourced Material
by Geological Source and Time Period
Table 2.8

## Discussion of Imported Artefacts

Obsidian is one of the most widely distributed long-distance trade goods. It appears during every time period from the Middle Formative to the Post-Classic. While contextual analysis is not traditionally included in obsidian studies, research conducted for this report reveals that this material was widely dispersed through a variety of contexts. It is likely that further analysis of this nature at other sites would reveal a similar dispersal pattern, suggesting that obsidian was not limited to a single segment of Maya communities nor to a specific role.

It should be noted that obsidian is not the only long-distance trade good present. Several other goods were know to have been traded over considerable distances including jade, *Spondylus* shells and other marine material, and during the Post-Classic, gold. A full discussion of this phenomenon may be found in Chapter 1, Section 2. While the focus of this research is on obsidian it is hoped that the methodology used to correlate material and context in an attempt to understand intra-site distribution patterns may be applied to other commodities.

**COASTAL SITES**

Three coastal sites were included in this report, Cerros, Moho Caye and Northern River Lagoon (Appendix I, figure 4). Through the years various different authors have focused on the role of coastal sites in Maya trade patterns. Consequently, an equally varied number of theories have been put forward. These theories include such ideas as the sites functioning as ports-of-trade (Andrews 1990; Chapman 1957; Rathje and Sabloff 1973; Sabloff and Rathje 1973), transhipment points (Guderjan *et al.* 1989; McKillop 1981, 1982; Hammond 1976), or way-stations tied to larger inland polities (Hammond 1976). As these different functions imply different levels of autonomy, it is natural to suspect that a correspondingly different resource pool would also be evident.

The most consequential form of trade centre for a coastal site would be that of a port-of-trade. Chapman defines a port-of-trade as a place "whose specific function was to serve as a meeting place of foreign traders" (1957:115). In order to facilitate the exchange of goods, a port-of-trade was deemed to require autonomy, as lack of neutrality, or the perception thereof, would cause "foreign traders and strangers [to] shy off and trade to suddenly dry up" (Chapman 1957:116). However, identifying sites that fit this criteria is difficult archaeologically, for, as Andrews suggests, "To define a site as a port-of-trade, one must have detailed knowledge of the subtleties of the economic and political structures and an understanding of how they operated" (Andrews 1990:165). It should be noted here that by Chapman's original definition, ports-of trade may, but do not have to be, associated with coasts or river trade routes (Chapman 1957). However, it is in this context that the term is most commonly used in Mesoamerican discussions.

Trans-shipment points are "nodes not only for coastal trade, but also as points from which long-distance goods were diverted to inland communities" (Andrews 1990:165). These sites are easier to identify archaeologically by the presence of diverse numbers of non-local materials. Often this term is used interchangeable with that of 'way-station' (Hammond 1976; Guderjan *et al.* 1989). Hammond described a way-station as a place "where the large coasting canoes would put in to off-load and take on goods from the mainland" (Hammond 1976:73). He goes on to suggest that "such stations would habitually be on small islands off the coast" and that such sites would likely be small and "unsuitability for settlement [of] any substantial scale" (Hammond 1976:73). While Hammond suggests that these sites would, like ports-of-trade, have been neutral areas, it is unlikely that locations which were "unsuitable for settlement" would have survived, much less thrived, regardless of the quantities of trade items shipped through the site. Andrews suggests that such sites were linked to larger centres located up-river (Andrews 1990).

A symbiotic relationship between coastal sites and a larger inland centre would seem reasonable, as the Belize coast is composed primarily of mangrove swamps, low-lying marshes, and multitudes of small inlets. Large centres located up-rivers, away from the coasts, had centres possessing greater protection from tropical storms and the subsequent flooding that occurs along the coast. It would also provide the population of these centres with a cultivatable territory. It is interesting to note that along the east coast of the Yucatan, coastal sites of sufficient size to suggest they were true ports-of-trade are those centres found in protected areas, such as Chetumal Bay and Cozumel Island, with terrain suitable for agriculture (cf. Sidrys 1983; Rathje and Sabloff 1973; Sabloff and Rathje 1975).

Virtually all trade centres, regardless of the exact nature of their political organisation, would demonstrate access to non-local resources, as the purpose of these sites, and how we recognise them, is by their role in redistributing trade goods. It is my opinion that the ability to differentiate between autonomous ports-of-trade or transhipment points can be found in the intra-site distribution pattern of these objects. Autonomous sites during the Classic Period would possess a hierarchical organisation that would demonstrate differential access to non-local status items, with larger quantities of such goods being found associated with elite residences or activities. Sites, which were politically dependent on a larger centre, would be less likely to exhibit differentiation in intra-site consumption patterns; and such non-local goods as might be present would be found in more general contexts spread throughout the settlement.

In an effort to investigate the possibility that such hypothesis might provide a valid means for determining possible differences in political organisation of coastal sites, the sites selected for this study were analysed as an independent group, as well as being incorporated into the general study of obsidian distribution and consumption patterns.

**Cerros**

One of the most well known Maya coastal sites is Cerros in the Corozal District, Belize (Appendix I, figure 4). The site was first noticed by Thomas Gann in 1900. In 1939 it was reported more fully by Gann and his wife (see Vail 1988). It was not until 1973 when Norman Hammond included it in his work on the archaeological sites in the Corozal District that the site appeared on a map (Hammond 1973). Sidrys (1983:160) also visited the site and conducted limited survey work there in 1973 and 1974. However, intensive investigations of the site were not conducted until 1974 when David Freidel secured a permit for the site and initiated a multi-year project designed to focus on as many aspects of the site as possible.

Located on a narrow spit of land named Lowy's Bight, with Laguna Seca on one side and Corozal Bay on the other, the site is situated on a promontory that afforded an unobstructed view of boats arriving along the coast (Freidel 1986; Vail 1988). Although the site may have been secure from surprise visitors its location left it dangerously exposed to the elements over the centuries. The majority of the site lies on land less than five metres above mean sea level and a section of the site has eroded into the sea (Freidel 1986; Vail 1988). Environmental reconstructions suggest that Corozal Bay was originally a closed lagoon, making the turbulent waters less dangerous to the site, and that the New River, which is now located to the south-west of the site, first met the bay in front of the site (Freidel and Scarborough 1982; Vail 1988). This maked the site secure from the threatening ocean while still providing access to marine and lacustrine resources.

Although originally identified by Gann as being a Post-Classic site (Gann 1900; c.f. Vail 1988), excavations rapidly revealed the error of this assessment. The site has a long history beginning in the early facet of the Late Formative Period and lasting until the Late Classic Period when the site was abandoned. It was reoccupied in the Late Post-Classic although construction and disruption to the original site was minimal. The resettlement is believed to be the result of merchant activity as the material found from this period includes rich caches of jade, obsidian, copper bells and a few artefacts of gold (Vail 1988). Freidel (1986) suggests that the people were utilising dugout canoes to exploit estuarine and coastal trade routes, and that they may have been travelling as far as the cayes to access resources.

Evidence of the earliest occupation comes from excavations in the central platform, that revealed a nucleated village dated to the early facet of the Late Formative Period (300 BC – 200 BC) (Cliff 1982). Evidence of increasing social differentiation, in terms of domestic architecture and items consumed in dedicatory caches, is apparent in the during this phase. During the succeeding phase the site orients itself around a central public area defined by small pyramids and residences that Cliff (1982) asserts were occupied almost exclusively by an elite. This site design is still visible today, with 5 large pyramidal structures complete with associated complexes. Four of these pyramid complexes are contained within the 4.4 hectares that form the site's core. The fifth complex is located to the south-east of the core but still within the perimeter of its as defined by the site's main canal.

Freidel (1986) implies that the construction of monumental architecture and the creation of iconographic friezes was part of a general transformation that occurred throughout the lowlands circa 50 BC. This transformation was the result of the elites from several communities acting in communication to form a set of identifiable conventions. The collaboration was accompanied by the exchange of goods as well as ideas (Mitchum 1994:4). This interaction and exchange with other areas of the Yucatan along with the Guatemalan Highlands can be seen in the increase of non-local items (Robertson and Freidel 1980:330). Freidel (1979) believes that the Cerros functioned primarily as a trade centre, and that the change from a small, egalitarian community into a socially stratified community with differential access to power and commodities was a direct result of Cerros' participation in the trading networks.

It was during the latter part of the Late Formative Period (50 BC – AD 200) that the nucleated settlement in the ceremonial core was abandoned in favour of dispersed villages (Vail 1988). Some occupation of the site centre was still present in the Early Classic, but at a greatly reduced scale. It has been suggested that exchange networks developed in the Formative Period were disrupted and that the managerial elites abandoned the area (Vail 1988). All occupation at the site ceased by the Late Classic.

Twenty-one pieces of obsidian from Cerros were analysed by NAA (Mitchum 1989). Two of these pieces were dated to the Post Classic period as such they were considered unsuitable for inclusion in this report, lowering the usable quantity of obsidian to 19 pieces (Table 2.9). All of these pieces were recovered from construction fill contexts and were therefor of indeterminate social context. However, Lewenstein notes that obsidian was found in many different contexts throughout the occupation of the site suggesting that these 19 pieces may represent a limited spectrum of consumption contexts (1987:156, 173, Tables 23 and 33)

| Date | ELC | SMJ | IXT | MX | Total |
|---|---|---|---|---|---|
| LF | 16 | 0 | 0 | 0 | 16 |
| EC | 1 | 0 | 1 | 0 | 2 |
| TC | 1 | 0 | 0 | 0 | 1 |
| Total | 18 | 0 | 1 | 0 | 19 |

Quantities of Cerros Obsidian
by Source and Time Period

Table 2.9

Establishing an accurate total for the complete obsidian collection of Cerros was exceedingly difficult. In 1991, Mitchum noted that the obsidian from this site equalled four per cent of the total chipped stone tool assemblage, a number given as "more than 15,000 pieces" (1991 :45). Lewenstein, although still vague, provides slightly more information stating that "approximately ten thousand chipped stone artefacts" were recovered from Cerros (1987 :137). Of these, Lewenstein analysed 4,814, six per cent of which were "grey obsidian blades" (1987 :138). From this, we can estimate that there were at least 288 pieces of obsidian recovered from excavations at Cerros (this being six per cent of the material Lewenstein analysed). If the relative proportion given for Lewenstein's study are projected onto her approximate total of 10,000 pieces, there may have been as many as 600 obsidian artefacts recovered, for the entire occupation history of the site. This latter number corresponds with Mitchum's four per cent of 15,000 pieces. Accepting the latter estimate for the quantity of obsidian, then the sourced material amounts to a mere three per cent of the obsidian collection. While this number is lower than one might wish, it is currently all that is available from this important site. Therefore we are compelled to include it, such as it is.

**Moho Caye**

This settlement is located on Moho Caye (Appendix I, figure 4), a small, flat island roughly 250 by 320 metres, four kilometres from the mouth of the Belize River (Healy *et al.* 1984; Vail 1988:43-44). Initially investigated by Gann in the early twentieth century, the site was classified as a fishing and hunting station, since the only evidence for occupation was middens consisting primarily of manatee remains (Vail 1988:43). Later surveys conducted at the island over the years revealed evidence of permanent occupation in the form of several low mounds (Vail 1988:43).

When surveyed in the late 1970s the island consisted predominately of mangrove swamps, with a small area at the northern point dry enough to allow for archaeological survey. Material collected from this area pointed to occupation during the Late Formative, Middle Classic, and Late Post Classic. Healy *et al.* (1984) identifies the Middle Classic as AD 400 to AD 700. The majority of activity at the island is believed to have occurred during this period; however, this belief may be flawed as rising sea levels may have flooded evidence. Thirteen features were identified belonging to the Middle Classic Period, including eight burials (Vail 1988 :44).

A total of 16 pieces of obsidian were reported for the site of Moho Caye. Three of these pieces were reported in Dreiss (1986), the remaining 13 in Healy, McKillop and Walsh's original (1984) report. Regardless of Healy *et al.* employing the temporal classification "Middle Classic" (AD 400 – AD 700), Dreiss and Brown (1989) assign all the pieces to the Early Classic Period. The rationale for this was that Healy assigned the pieces to the period immediately after the decline of Cerros (*c.f.* Dreiss 1988 :66). All the pieces used here follow Dreiss' lead (Table 2.1). Although all 16 pieces have both source and temporal contexts, only the original 13 artefacts from Healy *et al.*'s 1984 study were listed with archaeological contexts. Consequently, only these 13 artefacts were used in the larger tripartite analysis.

| Date | ELC | SMJ | IXT | MX | Total |
|---|---|---|---|---|---|
| EC | 13 | 0 | 3 | 0 | 16 |
| Total | 13 | 0 | 3 | 0 | 16 |

Quantities of Obsidian from Moho Caye
by Source and Time Period

Table 2.10

Non-local goods other than obsidian were recovered from the site. These include items produced from basalt, jade, chert, and granite (Vail 1988:44). It has been postulated that Moho Caye served a similar function as Northern River Lagoon, that of a transhipment location for trade goods (Healy *et al.* 1984; Vail 1988:44). Further investigation into the function of the island settlement was curtailed abruptly when the site was destroyed during construction of a marina by Belize Marine Enterprises Ltd. (Gutchen 1983:226).

**Northern River Lagoon**

Situated on the Belizean coast and located at the northern end of the lagoon, for which it was named (Appendix I, figure 4), the Northern River Lagoon site possesses a deep stratification dating to the Late and Terminal Classic periods. It has been suggested that this site was a specialised production area, manufacturing and distributing salt and salted fish to the nearby, inland sites (Mock 1997; Vail 1988:30-31; Valdez Jr. and Mock 1991). Northern River Lagoon was home to a full-time resident population that exhibited strong ties with the occupants of the site of Colha, roughly 15 kilometres inland. Although the main occupation of the residents at Northern River Lagoon may have been procurement of marine resources, it has also been suggested that the site served as a transhipment node for the importation of non-local resources for populations at Colha and the surrounding vicinity (Mock 1997; Vail 1988:30-31). This idea is supported by Mock, who describes the material culture of the site as including an "inordinate array of imported goods" (1997:165), far greater than would normally be expected for the size of the settlement.

Of these imported goods, obsidian artefacts are calculated as 38.6 per cent of stone tool assemblage (Mock 1997). Dreiss *et al.* (1993) included 20 pieces from Northern River Lagoon in their analysis of material from Colha. They found that of the 20 pieces examined, 17 pieces (85 per cent) were El Chayal in origin, while three pieces (15 per cent) were manufactured from Ixtepeque material (Table 2.11). All of these pieces dated to the Late/Terminal Classic Period and all were from domestic midden contexts.

| Date | ELC | SMJ | IXT | MX | Total |
|---|---|---|---|---|---|
| LC | 17 | 0 | 3 | 0 | 20 |
| Total | 17 | 0 | 3 | 0 | 20 |

Quantities of Obsidian from Northern River Lagoon by Source and Time Period

Table 2.11

**RIVER-ORIENTED SITES**

These sites are located adjacent to, or within easy distance of, navigable rivers or other water-ways that may have been used by the Maya for transporting goods (Appendix I, figure 4). While many sites in Northern Belize may fit this definition, material from only four sites was available for inclusion in this study. These sites and their related water-ways are as follows: Blue Creek at the nexus of the Rios Hondo, Bravo and Azul; Cuello along the New River; Nohmul along the Rio Hondo and Colha on the banks of Faber's Lagoon, also known as Cobweb Swamp.

Unlike coastal sites, river-based communities are unlikely to have had initial or direct contact with inter-regional merchants travelling along the coast. The "large coasting canoes" described by Hammond (1976:73) were likely too large to navigate the smaller, and often swampy, river areas. However, like the coastal sites, river-oriented centres may have served a similar role as 'transhipment' nexuses, transferring material borne by estuarine canoes to porters for portage overland to interior sites and vice versa. Andrews (1990), in his presentation of this type of settlement, recognises that the defining criteria for identifying a transhipment community is not a coastal location, but rather its ability to successfully redistribute material transported along different routes and from different regions. Hence, any site situated at the convergence of trade routes from which "long-distance goods were diverted to inland communities" may have the potential to be considered a transhipment point (Andrews 1990:165).

The most notable difference between coastal and river sites is size. The former sites are all relatively small, nucleated settlements, possessing a few ceremonial structures constructed close to each other in a single precinct. Sites located up-river, while not approaching the size of the Peten sites, tend to be considerably larger than their coastal cousins. River-oriented sites often possess multiple plazas, with less nucleated ceremonial structures. It is possible that these differences in settlement size and design may directly reflect the nature of the sites' functions. Sites functioning as 'way-stations' for specific inland sites may not have needed ceremonial structures, or structures on the scale of the parent community. Instead, residents may have participated in the rituals at the inland site. Concurrently, the lack of autonomy may have constrained the building impulses of the coastal community members, restricting their access to both the wealth needed for such construction as well as the permission, or right to possess such status items. The site of Cerros is an aberration in this pattern. However, as it appears that Cerros was not tied to any specific inland site (Freidel 1977, 1978, 1986; Freidel and Schele 1988; Robertson and Freidel 1986), it may more appropriately fit into the same category of 'transhipment' community as the river sites (Guderjan *et al.* 1989; McKillop 1981, 1982; Hammond 1976), than that of 'way station' of other settlements (Hammond 1976).

As these different functions imply different levels of autonomy, it is natural to suspect that a correspondingly different resource pool would also be evident. However, both of these site types would have had access to a variety of non-local goods in varying degrees. Consequently, it may be of value to consider such factors as site size, the type and nature of construction, and other activities at each site as a means to further identify areas of greater or lesser autonomy. As estuarine sites possess, albeit on a smaller scale, many of the same features of the larger, Central Peten sites (monumental architecture, elite tombs, caches, etc.), it is probable that these river-based sites exercised greater autonomy than the coastal way-stations. A comparison of the nature of the obsidian contexts and the activities which created them may help in elucidating the possible role of each site in the political and trading network of the region. Such a study is included in the next chapter of this report.

**Blue Creek**

Although Blue Creek is considered to be a river-oriented site, the Blue Creek Ruin was considered at the start of the discussion of sites. As the date models utilised in this study were created primarily from the analysis of the Blue Creek obsidian collection, it was deemed more appropriate to place this site first. Consequently, the full discussion of this site may be found at the start of this section.

**Cuello**

Located approximately 17 kilometres up the New River (Appendix I, figure 4) behind the modern Cuello Brother's Rum Distillery, for which it was named, the site was discovered during the British Museum-Corozal Project's 1973 regional survey (Hammond 1972; Pring and Walton 1976). Despite conflicts over the initial dating of the site

(Andrews and Hammond 1990), Cuello is one of the earliest settlements known in Northern Belize (Hammond 1977; Hammond *et al.* 1979; Kosakowsky 1987). The site, stretching approximately 1250 metres north to south and 600 metres east to west, consists of several platforms with numerous small structures on each (Hammond 1991a). Over 200 structures dating from the Early Formative through the Early Classic were identified (Gerhardt 1988; Gerhardt and Hammond 1991; Hammond 1991a).

A total of 367 pieces of obsidian were reported from the site of Cuello (Johnson 1976). Of these, 253 pieces were available for analysis, and 109 were sent for PIXE testing (Table 2.12) (Hammond 1991b; Johnson 1976). It should be noted that the inclusion of a source entitled "Other" was necessary, as the Cuello collection was not tested for San Martin Jilotepeque material, only for El Chayal and Ixtepeque sources (Hammond 1991b).

Thirty-eight pieces, all dated to the Late Formative Period, were identified as originating within the vague area entitled "settlement area" (Hammond 1991b). The remaining 71 pieces (Table 2.13) came from the Platform 34 excavations.

| Date | ELC | IXT | OTH | MX | Total |
|---|---|---|---|---|---|
| LF | 36 | 36 | 21 | 0 | 93 |
| EC | 9 | 6 | 1 | 0 | 16 |
| | 45 | 42 | 22 | 0 | 109 |

Total Analysed Obsidian From Cuello Identified by Source and Time Period

Table 2.12

| Date | ELC | SMJ | IXT | OTH | MX | Total |
|---|---|---|---|---|---|---|
| LF | 16 | 0 | 26 | 13 | 0 | 55 |
| EC | 9 | 0 | 6 | 1 | 0 | 16 |
| | 25 | 0 | 32 | 14 | 0 | 71 |

Obsidian from Platform 34, Cuello, Identified by Source and Time Period

Table 2.13

Platform 34 was initially constructed between 400 BC and 300 BC over a series of earlier, low, plastered platforms set around a patio (Hammond 1991b). These earlier domestic platforms were buried by rubble created from the destruction of nearby structures (Hammond 1991b). The most notable building on Platform 34 is Structure 35. This is a small, multi-phase pyramid, with the interior structure dating to the Terminal Formative Period, and the ultimate construction to the Early Classic.

All 109 pieces analysed were deemed suitable for inclusion into analysis of material by time period and analysis by social context. However, as no archaeological context could be determined for the 38 pieces from the settlement area, they were not included in the analysis of material by archaeological or social context.

**Colha**

Located on Lopez/Quashie Banner Creek at the southern edge of Faber's Lagoon, also know as Cobweb Swamp (Appendix I, figure 4), Colha is a small to medium-sized site possessing several small courtyard groups (Dreiss *et al.* 1993; Hester 1982). Settled in the Middle Formative, the site became an important tool production locale during the Late Formative, with 22 workshops dated to this period (Hester 1982). A population boom accompanied this rise to prominence in the Late Formative and construction at the site increased accordingly, with the establishment of monumental architecture (Eaton 1982). Although the prosperity of the lithic workshops is partly due to the fortuitous location of the site in the chert-bearing zone in Northern Belize, it has also been attributed to the development of intensive agricultural practices in the surrounding region during the Formative Period (Hester 1982), as these practices profit by the use of stone tools.

Colha possesses at least two main plazas and several smaller courtyard complexes, composed of five pyramidal structures and a variety of range structures (Eaton 1982). Since 1979, a variety of locations around the site have been excavated by the Colha Project under the direction of Thomas R. Hester (Dreiss *et al.* 1993). A total of 3,786 pieces of obsidian were recovered from a variety of contexts around the site (Dreiss *et al.* 1993). Several studies have been conducted over the years to identify the sources of the obsidian artefacts (Dreiss *et al.* 1993; Hester and Michel 1980). These studies include one of 14 pieces by Hester and Michel in 1980, another of 48 pieces by Dreiss in 1986, and a third by Dries *et al.* in 1993 that consisted of 180 artefacts from Colha and 20 from Northern River Lagoon.

Hester and Michel (1980) analysed 14 pieces by XRF at the Lawrence Berkeley Laboratory. Of these, only six fall within the temporal framework acceptable for this study (Table 2.14); the remaining eight were dated to the Post-Classic Period. Although this article provided the context of material examined, no attempt was made to discuss the distribution of material in this regard.

Dreiss' report on 48 pieces (1989) spanned the Middle Formative to the Post Classic (Table 2.15). These pieces were also analysed by XRF at the Lawrence Berkeley Laboratory. Although the majority of pieces were suitable for inclusion into the general analysis of material by time period, the lack of contextual data precluded their inclusion in the studies on contextual distribution.

| Date | ELC | SMJ | IXT | Total |
|---|---|---|---|---|
| LF | 0 | 1 | 4 | 5 |
| LC | 0 | 0 | 1 | 1 |
| Total | 0 | 0 | 0 | 6 |

Colha Obsidian Analysed
by Hester and Michel

Table 2.14

Dreiss *et al.*'s work (1993) is the most extensive of the Colha obsidian studies. This study included 180 items from Colha. One piece was discovered to be dark chert and subsequently was discounted, reducing the total of analysed obsidian to 179 pieces (Table 2.16). The pieces were dated by their associated ceramic assemblages and all were found to be in Tepeu contexts.

All 179 pieces were listed by operation designation, with information regarding the social status of the area's occupants also being provided. Only 99 pieces contained the necessary information to allow the objects to be securely assigned to an archaeological context. The article included an analysis of the material by context and status; however the results of this study were unsatisfactory. It is my opinion that the results were hampered by the lack of temporal diversity. Slight differences were noted between the distribution of El Chayal and Ixtepeque material in regards to ceremonial and elite contexts, leading Dreiss *et al.* to suggest that "El Chayal is more likely to be associated with elite activities" (1993 :279). Further evidence supporting this 'suggestion' is presented in the following chapter.

| Date | ELC | SMJ | IXT | Total |
|---|---|---|---|---|
| MF | 0 | 7 | 0 | 7 |
| LF | 4 | 4 | 6 | 14 |
| EC | 1 | 0 | 0 | 1 |
| LC | 4 | 0 | 5 | 9 |
| TC | 3 | 0 | 0 | 3 |
| PC | 1 | 0 | 13 | 14 |
| Total | 13 | 11 | 24 | 48 |

Colha Obsidian
from Dreiss' 1988 Study

Table 2.15

Due to discrepancies in the reporting process and the elimination of material falling beyond the temporal framework of this analysis, the quantities of Colha obsidian included for analysis in this report varies between tests. The total amount of material suitable for source/time distribution studies was 219 pieces (Table 2.17). The amount of material available for archaeological contextual studies was 105 objects (six from Hester and Michel's study, plus 99 pieces from Dreiss' 1993 report), and 179 pieces, all from Dreiss' 1993 work, were suitable for sociological studies.

| Date | ELC | SMJ | IXT | Total |
|---|---|---|---|---|
| LC | 27 | 1 | 50 | 78 |
| LC/TC | 51 | - | 28 | 79 |
| TC | 8 | 1 | 13 | 22 |
| Total | 86 | 2 | 91 | 179 |

Colha Obsidian Counts
From Dreiss *et al.*'s 1993 Study

Table 2.16

| Date | ELC | SMJ | IXT | Total |
|---|---|---|---|---|
| MF | 0 | 7 | 0 | 7 |
| LF | 4 | 5 | 10 | 19 |
| EC | 1 | 0 | 0 | 1 |
| LC | 31 | 1 | 56 | 88 |
| LC/TC | 51 | - | 28 | 79 |
| TC | 11 | 1 | 13 | 22 |
| Total | 98 | 14 | 107 | 219 |

Total Colha Obsidian
With Source and Temporal Contexts

Table 2.17

**Nohmul**

Folds in the limestone shelf of the Northern Belize Coastal Plain created a series of ridges running roughly north-south across the plain (Wright *et al.* 1959). Nohmul, like Cuello, Ka'Kabish and other sites in the region, was constructed on one of these ridges. With the Rio Hondo to the west and Pulltrouser Swamp to the east, the San Pablo ridge forms the modern boundary between the Corozal and Orange Walk Districts in Northern Belize. Nohmul is located on this ridge (Appendix I, figure 4), approximately 50 kilometres upstream from the mouth of the Rio Hondo (Hammond 1989). Covering an area roughly 20 square kilometres, the site consists of two main plazas and several outlying areas. Two other small sites in the area, San Victor to the north and San Luis to the south, may be associated complexes (Hammond 1985).

Originally named Douglas after a nearby community, the site, like many others, was poked and prodded by Gann in the early twentieth century and then again later in the 1930s with his wife (Hammond 1983). Gann renamed the site Nohmul, meaning "great mound" after the prominent pyramid on the Great Plaza (Hammond 1983). During the 1940s the site was quarried illegally by the Ministry of

Transportation for road fill, at which time three burials were uncovered and severely looted (Hammond 1983). The site was excavated by the British Museum-Cambridge University Corozal Project under the direction of Norman Hammond in 1973, 1974 and again in 1978 (Hammond 1973, 1983, 1985).

While several areas around the site were investigated by Hammond, the area of interest to this study is the Platform 137 Group. This platform was extensively excavated, revealing a long and complex pattern of occupation with Terminal Classic/Early Post Classic walled structures of Yucatecan style overlaying earlier, more traditional Classic Period structures (Hammond 1983). Forty-nine pieces of obsidian were recovered from the construction fill that formed the platform. These pieces were tested by NAA at the Brookhaven National Laboratory (Hammond *et al.* 1984). They span the Late Formative through the Terminal Classic Periods, with the majority of material coming from the later periods (Table 2.18). This is consistent with what Hammond describes as a "population maxima" in the Late Classic Period (1983:247).

| Date | ELC | SMJ | IXT | Total |
|---|---|---|---|---|
| LF | 1 | 0 | 0 | 1 |
| EC | 1 | 0 | 0 | 1 |
| LC | 8 | 0 | 19 | 27 |
| TC | 4 | 0 | 16 | 20 |
| Total | 25 | 14 | 35 | 49 |

Obsidian from Nohmul by Source and Time Period

Table 2.18

### INLAND SITES

These sites do not constitute a recognised region or area within the Maya world. The term is utilised here to distinguish specific sites by virtue of their location. Inland or middle sites are centres which are situated away from obvious trade or communication routes, and whose access to such routes may have been inhibited by other sites. This is not to exclude the possibility that these sites were part of an overland exchange network. Rather, it is to investigate the potential of using obsidian distribution in determining non-estuarine commercial routes that these sites are separated. Three sites were selected for this study: Ka'Kabish, Kichpanha, and Becan. While there are undoubtedly a number of other sites that may fit this category the selection of sites was restricted by the lack of available obsidian information.

#### Ka'Kabish

While known for many years, the site of Ka'Kabish has received little attention. Although Ka'Kabish was visited by David Pendergast while he was working at the nearby site of Lamanai, the first archaeological work to be conducted at the site was the mapping and surveying of the site by Thomas Guderjan in 1995 (Guderjan 1996). Despite having borne the name "Ka'Kabish" for several decades, ascribed by the local populace, the origin of the appellation is unclear. Barrera Vasquez's *Diccionario Maya* provides the following definition for the name:

> **KA'KAB-IS-AX**
> 10: (toponímico); **ka'kab:** aldea, asiento de población, tierra alta y fuerte; **is:** *Ipomoea batatas, Lam:* camote + **ax:** verruga; nombre de unas ruinas arqueológicas localizadas cerca de **Numk'ini (Nun k'ini,** Campeche).
> (Barrera Vasquez 1995)
>
> (10: (place-name); **ka'kab:** small village or hamlet, seat of population, high and strong land; **is:** *Sweet potato, LAm:* sweet potato + **ax:** wart; name of an archaeological ruin near Numk'ini, Campeche.
> (translation Haines)

It is possible the etymon of the current name may be Ka'Kabax. The location and description of the site corresponds closely to the definition of this word; the ruin is located on a section of *"tierra alta y fuerte"*, was once a village, and is an archaeological ruin. As the 'x' in Maya is pronounced as a "sh" sound (Gates 1978 :vi), it is possible that this spelling may have gradually mutated, or been anglicised into the current Ka'Kabish. It is also possible that the '-is', meaning sweet potato or Morning Glory, may be the correct ending. The site may have been named for plants growing nearby although this author has never seen any such plants. However, it should be noted that these are merely speculation, as the origin of the name has resisted attempts to be traced.

Like many other sites in northern Belize, Ka'Kabish was constructed on one of the many limestone ridges scattered across this area of the Coastal Plain (Hammond 1983). The road between San Filipe Town and Indian Church bisects the core area of the ruin. The construction and maintenance of this road has resulted in the site being subject to a wide spectrum of destructive activity, including quarrying operations by the Ministry of Transportation, as well as the usual activities of looters. Ka'Kabish unfortunately suffered a fate similar to that of Nohmul mentioned above; the end of at least one structure has been completely removed by the Ministry of Transportation, and a large chunk has been taken out of one plaza to supply marl and limestone for road fill. The surrounding area is currently cultivated as small milpas by the occupants of San Filipe and Indian Church.

The site's core area consists of two plazas containing 26 buildings. Several other structures were noted in the surrounding areas but are currently unmapped. Those structures, 1 to 15, to the south of the San Filipe/Indian Church Road were identified by Guderjan (1996) as belonging to the "Main Plaza". It is in this plaza, between Structures 3 and 4, that the Ministry of Transportation sought quarrying materials. North of the road are structures 16 to 27. Structure 25 is on private land and was not mapped (Guderjan 1996).

Looted tombs were discovered in several structures (Guderjan 1996). Obsidian was noted occurring both in the exposed shafts above the tombs and scattered across the floors. The obsidian collected came from four areas: Collection Area-A, Collection Area-H, Structure 17 (Tomb 1), and Structure 22 (Tomb 4). Both of these tombs were vaulted and the walls of Tomb 4 exhibited evidence of having once been painted red. The east wall of Tomb 4 retains evidence of images, possibly fragments of glyphs, painted a dark red on a lighter red backdrop, although they have been too badly damaged to allow for a positive identification (Martin personal communication).

A total of 836 pieces were collected from these four locations, amounting to a total of 791.6 grams (Table 2.19). Of these 836 pieces, 86 (10.3 per cent) were visually identified as having been manufactured from El Chayal material. This was the only source noted during inspection of the collection. Testing of the material by NAA was unfeasible, as it was not possible to export the material.

| Location | Quantity | Weight g. |
|---|---|---|
| Collection Area-A | 137 | 87.8 |
| Collection Area-H | 216 | 99.6 |
| Structure 17, Tomb 1 | 431 | 507.4 |
| Structure 22, Tomb 4 | 53 | 96.8 |
| Total | 836 | 791.6 g |

Ka'Kabish Obsidian
by Piece and Weight

Table 2.19

| Date | ELC | SMJ | IXT | MX | Total |
|---|---|---|---|---|---|
| LC | 86 | 0 | 0 | 0 | 86 |
| Total | 86 | 0 | 0 | 0 | 86 |

Material from Ka'Kabish
by Source and Time Period

Table 2.20

Material collected from these areas included ceramic shards; it is with these that the rough date of "Late Classic" is assigned to the construction of the tombs, and by association, to the obsidian deposits. The exposed area of construction shaft in Tomb 4 revealed layers of obsidian, shells, chert flakes and carbon alternating with layers of stone construction fill. All of the obsidian is considered to be part of the Tomb Shaft ritual, and as such was ascribed a social context of Ritual/Ceremonial 2 (RC2) (see previous section for complete definitions).

### Kichpanha

Located on a limestone ridge running roughly north and east, the site of Kichpanha is a small site, encompassing approximately 4.5 square kilometres. The ridge upon which the site is situated is near the northern limit of the chert-bearing zone in Northern Belize (Gibson 1982). To the south-west of the ridge is Kate's Lagoon, and approximately 12 kilometres to the south-east lies the site of Colha (Gibson 1982). Although situated on the banks of a lagoon, this site is considered an "inland" site by virtue of the isolated nature of the lagoon. Kate's Lagoon is not connected to any streams or other navigable water-way.
Initially identified by Hammond (1973) from bulldozer cuts, it was speculated that the site was a Formative settlement and lacked an Early Classic occupation component. Later excavations at the site revealed occupation spanning from the Middle Formative through to the Early Post-Classic (Gibson 1982). The site possesses two courtyards with monumental structures, between which are numerous smaller structures (Gibson 1982).

A total of seven pieces of obsidian were reported as having been subjected to NAA testing (Dreiss 1986, Dreiss and Brown 1989). These pieces date to the Formative and the Late Classic periods (Table 2.12).

| Date | ELC | SMJ | IXT | Total |
|---|---|---|---|---|
| MF | 1 | 1 | 2 | 4 |
| LF | 1 | 0 | 1 | 2 |
| LC | 1 | 0 | 0 | 1 |
| Total | 3 | 1 | 3 | 7 |

NAA Sourced Material from Kichpanha

Table 2.21

### Becan

Located roughly 150 kilometres north of Tikal, Becan is situated in the heart of the Rio Bec Region. The site was discovered by Karl Ruppert and John Denison in 1934, and consequently the honour of naming the new-found ruin fell to them. They chose to name the site after its most distinct characteristic, a fosse surrounding the site. Barrera, using the Maya spelling of Bekan gives the following definition:

> **BEKAN**
> 2: [foso, cavidad, cava, barranca]; **u bekanil pa':** cava de fortaleza 3: barranca o quebrada; **u bekanil paa', u batkabil u pach paa':** cava o fosa de fortaleza

(Barrera Vasquez 1995)

[trench, cavity, dig, gully]: **u bekanil pa':** excavated fortification 3: gully or ravine **u bekanil paa', u batkabil u pach paa':** fortified pit
(translation Haines)

Sharer provides a more concise, although slightly different interpretation: "ditch filled with water" (1994 :200). Sharer's definition implies that the fosse originally contained water, an assumption that is not mentioned anywhere else in the literature on Becan (Rovner 1989; Sharer 1994:200-203). Water-filled or not, the fosse formed an impressive fortification, roughly 16 metres wide and originally five metres deep, the defensive nature of the construction was expanded upon with the erecting of an earthen motte, somewhat fancifully described as a rampart (Sharer 1994:200), 5 metres tall. Both the fosse and earthwork enclose the core area, approximately 114 hectares.

Settled in the latter part of the Middle Formative (c. BC 550), the site flourished into the beginning of the Early Classic, when the defensive constructions were initiated (Rovner 1989; Sharer 1994:200-203). Shortly after, Becan's population declined for reasons that are not clear. In the Late Classic, the site was revitalised with a population boom and corresponding flurry of building activity (Rovner 1989; Sharer 1994:200-203). It is to this period that the distinctive Rio Bec twin-tower architecture originates (Sharer 1994). This renaissance was to be short-lived, ceasing in the early ninth century. Ceramic evidence documents a significant demographic shift during that century, with people migrating into the area from the Northern Yucatan (Sharer 1994:200-203). Despite this influx, the population of the area continued to decline until the region was abandoned in the Early Post Classic.

Excavations at the site began in 1969 and continued for three years under the direction of E. Wyllys Andrews IV with the support of Tulane University and the National Geographic Society (Sharer 1994). A total of 194 pieces of obsidian were collected during these excavations, and all but one, a large bifacial point, was analysed to determined the origin of the pieces (Rovner 1989). The pieces were subjected to PIXE tests at the University of Western Michigan's Department of Physics (Rovner 1989). Forty-eight of these pieces were discovered in mixed Terminal/Post-Classic deposits and hence were deemed unsuitable for inclusion in this study. Another eight pieces from temporally indeterminate contexts were also excluded, lowering the potential usable sample to 137 pieces (Table 2.22).

The majority of material (67.9 per cen*t*) was collected from Early Classic deposits. Less than 10 per cent of the collection (9.5 per cent) belonged to the Late Formative Period. Late Classic material was slightly better represented, consisting of 21.9 per cent of the sample, while the Terminal Classic, represented by only one item, accounted for a mere 0.7 per cent of the collection. The distribution patterns of sources utilised through these periods are not particularly surprising. Reliance upon the El Chayal source is dominant through both the Late Formative and Classic periods, with a greater dependency in the Classic Periods. The lack of Ixtepeque material in the Terminal Classic sample is not noteworthy as that collection is too small to reveal any pattern. What is peculiar is the high percentage of Mexican material present in the Early Classic Period. This material comprises 14 per cent of the sample, significantly higher than many of the other sites in the surrounding region (this study, *c.f.* also Dreiss 1989). Only the large sites in the Central Peten have Early Classic collections with greater quantities of Mexican material; obsidian from Mexican sources comprises 25.9 per cent of the Tikal collection and 81 per cent of the El Mirador collection (this study). Four of the Becan pieces were projectile points recovered from the fosse. It has been hypothesised that these pieces were deposited in the fosse after the construction of the fortifications and may have been used in battle (Rovner 1989:370).

| Date | ELC | SMJ | IXT | MX | Total |
|---|---|---|---|---|---|
| LF | 8 | 3 | 2 | - | 13 |
| EC | 77 | 1 | 2 | 13 | 93 |
| LC | 26 | - | 4 | - | 30 |
| LC/TC | 1 | - | - | - | 1 |
| Total | 112 | 4 | 8 | 13 | 137 |

Quantities of Material from Becan
by Source and Time Period

Table 2.22

While the 137 sourced pieces from the Formative and Classic Periods were suitable for inclusion in the general investigation of the source/time distribution pattern, they were not all suitable for inclusion in the more complicated contextual analyses. Only 76 pieces, all from the Early Classic, were accompanied by contextual information. Twenty-two pieces were found in construction fill, El Chayal material accounted for 18 of these pieces, and Ixtepeque and Mexican sources were represented with 2 objects apiece. The remaining 54 objects were discovered in cache context, and, as with the construction fill, El Chayal material composed the majority of the collection, accounting for 51 pieces. Objects manufactured from Mexican material made up the last three objects.

## Central Peten Sites

Loosely defined as the 'central lowlands', the Peten is a vast area carved out of the middle of the Yucatan. Lying at the heart of the Lowland Maya realm, the Peten consists of extensive tracts of rainforest interspersed with small rivers, lakes, and a series of seasonal *bajos* or *akalcheob* that range between low, undulating ridges formed from the Cenozoic limestone shelf. The rainforest is subject to irregular rainfall, resulting in long dry spells. While the arboreal vegetation may reach as high as 50 metres, the canopy roof is lower, consisting of ramon, sapodilla, fig, and palm, as well as many other plants.

This area was once considered by Morley to be the "Old Empire" of the Maya (Morley 1946). As Formative occupations have been found in other regions, the idea that the Central Peten was the cultural womb from which such aspects as political, social, and economic organisation were conceived before being spread to other areas in the Lowlands has largely fallen by the wayside (Rathje 1971, 1972). This area does contain some of the most extensive centres, along with some of the densest occupation (Hansen 1990; Puleston 1983; Tourtellot 1988, 1990). The following section examines the obsidian from two centres and one regional study. The two sites examined are Tikal and El Mirador, while the regional study consists of a series of excavations into the settlement basins of six lakes. This latter project was part of the Central Peten Historical Ecology Project, here identified as the "Central Peten Lakes".

Both Tikal and El Mirador are large centres with multiple plazas, each possessing ceremonial structures. Occupation at both sites was initiated in the Formative Period, and monumental constructions at both sites date to this period. Previously, it had been thought that Tikal was the oldest site to possess monumental ceremonial architecture on a grand scale. The prominence of this site is reflected in its material culture (Moholy-Nagy 1994). Several theories have been put forth linking Tikal to inter-regional commerce with Teotihuacan and Kaminaljuyu in a grand *menage à trois* of commercial dominance (Arnaud 1990). Other more restrained theories speculate on Tikal's role as a redistributive nexus, controlling the influx and distribution of non-local resources throughout the region. These studies assume the supremacy of Tikal based on its size and early history of political organisation. However, recent studies at the site of El Mirador suggest that this site, and the nearby centre of Nakbe, may have been large ceremonial centres before Tikal achieved such eminence (Hansen 1990; Sharer 1994). Explorations at these sites are not as complete as those at Tikal, one of the most extensively investigated sites in the Maya realm, so the impact of these new sites on hypotheses regarding inter-regional exchange activity for the Peten is uncertain. An examination of the contextual uses of these materials may help clarify discrepancies in trade access between sites within the region as well as with other regions.

The sites from the Central Peten Lakes, being smaller, domestic oriented occupations, were included in this analysis to provide a more holistic understanding of the trade patterns within the region. It is clear from the nearly 300 pieces of obsidian recovered from the Central Peten Lakes excavations that larger ceremonial centres in the region were not the sole beneficiaries of long-distance trade activities.

### Central Peten Lakes

Northern Guatemala's Department of Peten is a region of tropical rainforest with undulating karst terrain and numerous small rivers connecting a series of small lakes. The settlement basins of six of these lakes formed the focus of the Central Peten Historical Ecology Project (CPHEP). This was a multidisciplinary endeavour designed to investigate interactions between human beings and lacustrine environments, with a focus on pre-historic Maya occupation (Rice 1984; Rice, *et al.* 1985). The lakes chosen for this study were Yaxha, Sacnab, Macanche, Salpeten, Quexil, and Petenxil (Rice 1985). The area has a long history of occupation beginning in the Middle Formative and lasting until the Spanish Conquest in AD 1525. The sites investigated were primarily domestic and rural in nature.

Excavations were conducted on 355 structures and plazas in the basins around these lakes (Rice 1984). Construction fill from these excavations yielded a total of 826 pieces of obsidian, 294 pieces, roughly 35 per cent of the collection, was tested to identify their source of origin. These pieces were found in structural contexts, a mixture of construction fill and reused midden material (Rice 1984, Rice, *et al.* 1985). The integration of these deposits into a single context is not unusual. However, as none of the material could be securely assigned to domestic midden contexts, the archaeological context of the objects was considered to be "construction fill". The rationale for this choice was that as objects in both construction fill and middens may be considered to have undergone a transformation from a systemic to archaeological context, probably through discard, and as the construction fill was the ultimate depositional location, assignation of the material to this context formed the most secure choice. Classification of the material to this context is supported by Rice, who stated that "most excavations were into construction fill and thus virtually all the analysed obsidians come from construction fill contexts" (1984). Rice minimised potential temporal problems caused by mixed deposits by selecting only obsidians from unmixed temporal contexts. Dates for the deposits were established by ceramic chronology using the latest ceramics from each sample.

Those items chosen for analysis span all occupation levels of the site, from the Middle Formative through to the Post Classic (Table 2.23). The material from the Post Classic was not considered in this analysis as it falls beyond the defined temporal parameters. Of the remaining 198 pieces, 31.8 per cent belonged to the Middle Formative Period; this is the largest collection from this site. The Late Formative sample is smaller, consisting of only 17.2 per cent of the tested collection, which is followed by an even smaller 12.6 per cent of Early Classic material. The Late Classic sample at 30.3 per cent is comparable to the Middle Formative collection. Terminal Classic objects account for a mere 8.1 per cent of the sample.

| Date | ELC | SMJ | IXT | OTH | Total |
|---|---|---|---|---|---|
| MF | 11 | 47 | 3 | 2 | 63 |
| LF | 8 | 22 | 4 | - | 34 |
| EC | 18 | 6 | - | 1 | 25 |
| LC | 40 | 13 | 4 | 3 | 60 |
| TC | 12 | 1 | - | 3 | 16 |
| PC | 27 | 20 | 46 | 3 | 96 |
| Total | 116 | 109 | 57 | 12 | 294 |

Obsidian from the Central Peten Lakes Project by Source and Time

Table 2.23

Sourcing of the material was accomplished primarily through XRF at the Lawrence Berkeley Laboratory in California. Thirty-five pieces required further abbreviated NAA testing, while three objects could only be assigned to source through a complete NAA examination. Patterns for the distribution of the material in each period roughly matches that documented for the other areas discussed. San Martin Jilotepeque is the dominant source in the Formative Period, comprising 74.6 per cent of the Middle Formative and 64.7 per cent of the Late Formative collections. The supply of obsidian shifts in favour of El Chayal material during the Classic Period, when this source represents 72 per cent of the Early Classic sample and 66.7 per cent of the Late Classic collection. The dominance of El Chayal material in the collection persists into, and increases in, the Terminal Classic, when it comprises 75 per cent of the material tested. Oddly, no Ixtepeque material appears in this sample, although in the preceding Late Classic Period it represents 6.7 per cent of the collection. The reduced quantity of Ixtepeque material in the Terminal Classic Period is the only distinct aberration in what has become a familiar pattern of obsidian utilisation in the Maya area.

**El Mirador**

Located 105 kilometres north-east of Tikal and a scant seven kilometres south of the Mexican border, El Mirador is in a remote section of the Northern Department of Peten, Guatemala (Howell 1989; Matheny *et al.* 1980). Named by local Chicleros to mean "the lookout" (Howell 1989), the site oversees a section of the Peten characterised by well drained uplands and poorly drained seasonal bajos. El Mirador was constructed in an upland area marked by a steep escarpment and with a large bajo to the west.

First discovered in the 1920s by two explorers working for the Shufeldt Co., the site was later host, albeit briefly, to an expedition from the Carnegie Institution of Washington who stayed a few hours (Ruppert and Denison 1943:49 in Mathney *et al.* 1980). The site was ignored by archaeologists until the 1960s, when Ian Graham visited the site and produced the first known map of the ruins (Graham 1967, Matheny, *et al.* 1980). Graham observed a distinct lack of fine cut masonry at the site and concluded that, despite its size, the site was earlier than Tikal.

In 1978, Bruce Dahlin began the Project Acalaches to investigate the utilisation of the seasonal swamps around the site of El Mirador. The following year, Dahlin and Ray Matheny from the New World Archaeological Foundation at Brigham Young University formed the El Mirador Project that from 1979 to 1983, investigated the site and its surrounding environment. This project identified several sizeable reservoirs and bajos around the site and a series of long causeways or *sacbeob* and footpaths radiating out from the site centre, crossing the wet areas. One *sacbe* connects El Mirador with the site of Nakbe (Howell 1989). Matheny, Hansen and Gurr (1980) suggest that these paths may be eco-barriers designed to enhance water catchment.

Three principle architectural complexes were identified at the site: El Tigre, Dante, and Tres Mariás. Obsidian used in this study was recovered from El Tigre, the westernmost complex. This complex is dominated by a pyramid 55 metre high and approximately 150 metre wide at its base. This imposing structure is oriented southward and is associated with the Central Acropolis, a complex of structures 335 metre long and 110 metre wide, an area that Howell (1989) refers to as "the downtown". Hansen (1990:209) places the earliest occupation at the site at 400 BC, with El Tigre complex floors being laid circa 300 BC. The earliest architecture at the site is believed to have been simple residential structures constructed with low stone walls, clay or plaster floors, and perishable superstructures. Modest ceremonial structure of similar composition were also constructed around this time, with the more massive construction projected being initiated between 150 BC and 1 BC (Hansen 1990:209-210).

The majority of the obsidian used in this analysis comes either directly from El Tigre complex, or from smaller structures located adjacent to the complex. The three excavations for which enough information could be collected to warrant their inclusion are Operations 26, 32, and 36 (Hansen 1990). Operation 26 was situated in the Structure 34 Complex, a large T-shaped platform on the southern border of El Tigre platform. The building is approximately 9 metre high with an east-west axis of 71 metre at the base and is oriented with the frontal staircase facing the El Tigre pyramid (Hansen 1990). Objects from Operation 32 comes from two units, 32c and 32e, both were placed in the centre of El Tigre Platform and containing material identified by Hansen as "workshop debris" (1990:196). The upper platform and summit of El Tigre were classed as Operation 36. This is a large 30 metre high platform with three discrete structures arranged in a triadic or T-shaped configuration which has also been documented at the sites of Nakbe, Tintal, and Uaxactun (Hansen 1990:101). These structures, identified by Hansen (1990), are 4D3-1, the largest at roughly 23 metre high, and 4D3-2 and 4D3-3, both approximately 10.5 metres high.

| Date | ELC | SMJ | IXT | MX | Total |
|---|---|---|---|---|---|
| LF | 50 | 7 | 2 | - | 59 |
| EC | 3 | - | 1 | 13 | 17 |
| LC | 11 | 1 | - | 1 | 13 |
| Total | 64 | 8 | 3 | 14 | 89 |

Total Obsidian Reported from El Mirador (Nelson & Howard 1986; Hansen 1990)

Table 2.24

Obsidian from the site was reported in two different articles: Hansen's 1990 report, *Excavations in the Tigre Complex, El Mirador Peten, Guatemala*; and Nelson and Howard's more specialised report, *Trace Element Analysis of Obsidian Artifacts from El Mirador, Guatemala,* from 1986. Material from both reports was compared and where clear repetitions were present, the material is recorded here only once with reference made to both reports (Table 2.24).

Nelson and Howard's report includes material that was not suitable for inclusion in this study due to missing information. Of their 13 Late Classic pieces that were sourced, only five could be included in this report. The number of Early Classic artefacts suitable for this study was higher, with 10 of their 16 reported pieces being included. The Late Formative had the lowest amount of usable material with only nine of their 42 pieces having enough accompanying information to warrant their inclusion. These 24 pieces account for 33.8 per cent of the 71-piece sample that was reported by Nelson and Howard (1986). Due to the organisation of Hansen's report, the necessary information was available for all 17 Late Formative Pieces and 11 Early Classic pieces mentioned therein (Hansen 1990).

The combined collection includes 26 pieces from the Late Formative, 13 from the Early Classic, and 5 dated to the Late Classic Period (Table 2.25). The Late Formative pieces used in the contextual studies are divided between the El Chayal and San Martin Jilotepeque sources, with 22 pieces (84.6 per cent) coming from the former source and four pieces (15.4 per cent) belonging to the latter (Table 2.25). This closely approximates the distribution pattern for the total obsidian analysed from this period, where 84.7 per cent (n=50) was El Chayal in origin, 11.9 per cent (n=7) came from San Martin Jilotepeque and 3.4 per cent (n=2) was from Ixtepeque (Table 2.24).

All 13 of the Early Classic pieces were from Mexican sources (Pachuca n=1, Paredon n=1, Zinapecuaro n=1, Ucareo n=2, Zaragoza n=2, Otumba n=3). While at first glance this number may seem high due to material not being included, of the 17 total reported pieces for the Early Classic Period only 4 (El Chayal n=3, Ixtepeque n=1) or less than 25 per cent were from Guatemalan sources (Table 2.24).

| Date | ELC | SMJ | IXT | MX | Total |
|---|---|---|---|---|---|
| LF | 22 | 4 | - | - | 26 |
| EC | - | - | - | 13 | 13 |
| LC | 4 | - | - | 1 | 5 |
| Total | 26 | 4 | - | 14 | 44 |

Total Obsidian Reported from El Mirador Suitable for Inclusion in Contextual Analyses

Table 2.25

Of the Late Classic pieces, four (80 per cent) were El Chayal in origin while one (20 per cent) came from Central Mexico (Table 2.25). This division reflects that present in the Late Classic collection as a whole, where 11 (84.6 per cent) of the artefacts were determined to be El Chayal, and San Martin Jilotepeque and Mexican sources yielded one piece (7.7 per cent) each (Table 2.24).

**Tikal**

One of the most famous, most visited, and longest excavated sites in the Maya realm is Tikal in the Department of Peten in Northern Guatemala. Tikal appears in the historical records of Guatemala as early as the eighteenth century (Coe 1965). Archaeological investigations at the site commenced in the early 1880's when Alfred Maudslay visited the site (Maudslay 1897). During the late nineteenth and early twentieth centuries, the site played host to a series of archaeological dignitaries including Teobert Maler in 1895 and 1904, Alfred Tozzer, Sylvanus Morley in 1914, 1921, 1922 and 1928, Edwin Shook in 1937 and 1942, and Tatiana Proskouriakoff in 1942 (Coe 1965). In 1956, the Tikal Project was created under the direction of Shook and backed by the University Museum of the University of Pennsylvania with the support of the Guatemalan government. This project was, and remains, one of the most extensive archaeological undertakings in the region. Running from 1956 until 1969, work was conducted excavating and consolidating the structures, as well as investigating the botanical, biological, geological and environmental aspects of the region (Coe 1965). In 1970, the Tikal Project was replaced by the Tikal National Park Archaeological Project which continued the work of excavating and consolidating structures and made the site accessible to visitors.

As Tikal is one of the largest and most extensively excavated Maya sites, it is not surprising to discover that the quantities of obsidian are correspondingly larger than collections from other sites. However, arriving at an accurate total for the obsidian from Tikal may prove impossible if, as Moholy-Nagy intimates, some obsidian was "encountered in the excavations but not collected or recorded" (Moholy-Nagy 1994:67). In her work on the material culture from Tikal, Moholy-Nagy analyses over 62,000 pieces of obsidian (1994), but notes that the total quantity of obsidian from Tikal may exceed 400,000 pieces (1994:72). An equally impressive number of papers have been produced over the years focusing on analysing the obsidian or discussing these analyses (Dreiss and Brown 1989; Moholy-Nagy 1975, 1976, 1984, 1989, 1994; Moholy-Nagy and Nelson 1987, 1990; Nelson 1985; Nelson *et al.* 1977, 1978; Stross, *et al.* 1971, 1968). Alas, although many of these are syntheses of material presented in earlier papers, few exhibit any data concordance. Consequently, data from only two reports was incorporated into this analysis. The material utilised was gleaned from Moholy-Nagy's 1975 report (Table 2.26) and Moholy-Nagy and Nelson's 1990 article (Table 2.27).

Material from the earlier paper represents the results from 203 pieces of obsidian. A combination of XRF and NAA was conducted on the pieces with eight pieces analysed by XRF at the University of California, Berkeley and 195 by NAA at the University of Michigan. Not all of these pieces were suitable for inclusion in this analysis; five pieces for which there was neither dates nor contextual information had to be discounted. Furthermore, Moholy-Nagy lists several pieces as being from Early/Late Classic deposits. For the purposes of this analysis, these items were considered Late Classic. None of these pieces included contextual information.

| Date | ELC | SMJ | IXT | MX | UNK | Total |
|---|---|---|---|---|---|---|
| LF | 7 | 27 | 15 | 2 | - | 50 |
| EC | 26 | 5 | - | 10 | 1 | 42 |
| EC/LC | 26 | 9 | 1 | 13 | 2 | 51 |
| LC | 38 | 2 | 2 | - | 2 | 44 |
| TC | 2 | - | 4 | 3 | 1 | 10 |
| ND | - | 1 | - | 4 | - | 5 |
| Total | 99 | 44 | 22 | 32 | 6 | 203 |

Sourced Material from Tikal By Time Period (Moholy-Nagy 1975)

Table 2.26

In their 1990 article, Moholy-Nagy and Nelson report on 30 artefacts (1990). One of these pieces (Table 2.27: italicised) was discovered not be obsidian, but possibly a tektite (1990:75), making the usable quantity of obsidian in this report 29 pieces. The pieces discussed were the same as those presented initially in a 1984 article (Moholy-Nagy 1984). Element analysis was conducted by XRF, then confirmed on a test sample from each group by NAA (Moholy-Nagy 1984).

The 1990 report was chosen over the initial paper as the later work included information regarding the social and archaeological contexts of the material. After sorting the data, it became apparent that 227 pieces (Table 2.28) were acceptable for analysis by source and temporal correlates, while 29 pieces (Table 2.29) could be consolidated into the more complete studies incorporating contextual information.

| Date | ELC | SMJ | MX | UNK | Total |
|---|---|---|---|---|---|
| LF | 2 | 5 | - | - | 7 |
| EC | 6 | - | 4 | 2 | 12 |
| EC/LC | - | - | 1 | - | 1 |
| LC | 6 | - | 2 | *1* | 9 |
| TC | 1 | - | - | - | 1 |
| Total | 15 | 5 | 7 | 3 | 30 |

Sourced Material from Tikal By Time Period (Moholy-Nagy and Nelson 1990)

Table 2.27

The material collected for source/temporal analysis contained almost equal amounts of Late Formative and Early Classic material (25.1 per cent and 23.8 per cent respectively). The majority of the material came from the Late Classic, which yielded 46.3 per cent of the material.

Terminal Classic objects accounting for the remaining 4.8 per cent. Within each time period, the material proved to be distributed along familiar patterns. San Martin Jilotepeque objects dominated in the Late Formative Period, accounting for 56.1 per cent of the sample, while items from Ixtepeque, El Chayal, and Central Mexico (26.3 per cent, 15.8 per cent and 3.5 per cent respectively) lagged significantly behind. In the Early Classic Period, the expected shift to reliance upon the El Chayal source occurred, with material from this source increasing to 59.3 per cent. Central Mexican pieces also increased, representing 25.9 per cent of the sample, while San Martin Jilotepeque material was reduced to 9.3 per cent. Several pieces were unknown and these make up 5.5 per cent of the Early Classic material. Dominance of the El Chayal material continued into the Late Classic, increasing to 76.2 per cent of the sample. Material is roughly equally distributed between the Central Mexican (12.4 per cent) and San Martin Jilotepeque sources (10.5 per cent), while Ixtepeque items accounted for a mere 2.9 per cent, less than the unknown material at 3.8 per cent. Terminal Classic objects show a slight shift toward Ixtepeque material (36.4 per cent), with Central Mexican and El Chayal pieces being equally represented at 27.3 per cent each, and unknown items accounting for 9.1 per cent. However, before too much emphasis is placed on this pattern of increased reliance on Ixtepeque material in the Terminal Period, the small number of items should be noted. This warning should be repeated for all the Tikal material, as it represents such a small portion of what may be one of the largest collections of obsidian reported.

| Date | ELC | SMJ | IXT | MX | UNK | Total |
|---|---|---|---|---|---|---|
| LF | 9 | 32 | 15 | 2 | - | 57 |
| EC | 32 | 5 | - | 14 | 3 | 54 |
| LC | 80 | 11 | 3 | 13 | 4 | 105 |
| TC | 3 | - | 4 | 3 | 1 | 11 |
| Total | 124 | 48 | 22 | 32 | 6 | 227 |

Sourced Material from Tikal By Time Period (Moholy-Nagy 1975, Moholy-Nagy and Nelson 1990)

Table 2.28

| Date | ELC | SMJ | IXT | MX | OTH | Total |
|---|---|---|---|---|---|---|
| LF | 2 | 5 | - | - | - | 7 |
| EC | 6 | - | - | 4 | 2 | 12 |
| LC | 6 | - | - | - | 3 | 9 |
| TC | 1 | - | - | - | - | 1 |
| Total | 15 | 5 | - | 4 | 5 | 29 |

Sourced Material from Tikal By Time Period Suitable for Inclusion in Contextual Analyses

Table 2.29

**Tikal/Yaxha Corridor**

Unlike the previous sites mentioned, this excavation does not focus on one particular site, but on a region between two sites, Tikal and Yaxha in the Central Peten. Information about Tikal may be found above. Yaxha is a smaller site located roughly 30 kilometres to the south-east of Tikal on the north Shore of Lake Yaxha. Visited by Maler in 1904, the site was not mapped until the Carnegie Institution of Washington explored the site in the 1930s (Sharer 1994:194). Work was also conducted more recently by Nicholas Helmuth in the 1970s (Sharer 1994:194). The site consists of a series of plazas, acropoli, and *sacbeob*.

By concentrating on the area between these two centres, the project was able to investigate settlement and residential locations away from core areas. In order to accomplish this study, the project established a base line roughly 28 kilometres long connecting the two centres and surveyed a transect 500 metres wide along this line. Twelve locations were excavated in this transect, all yielding obsidian artefacts. A total of 63 pieces of obsidian were recovered, all of which were analysed and identified to their original source. Testing of the material was conducted by the Lawrence Berkeley Laboratory using non-destructive XRF techniques. The obsidian was recovered from middens associated with the 12 residential units excavated. Dating of the material was accomplished through a joint process of ceramic correlation and obsidian hydration. The material was found to span all time periods from the Late Formative through to the Terminal Classic (Table 2.30). Seven artefacts were not assigned to a temporal context. These pieces were omitted from the analysis and Table 2.30 reflects their absence, making the usable total of artefacts from this site 56.

| Date | ELC | SMJ | IXT | MX | Total |
|---|---|---|---|---|---|
| LF | 3 | 10 | - | - | 13 |
| EC | 1 | 2 | - | 1 | 4 |
| C | 10 | 2 | - | - | 12 |
| LC | 24 | 1 | - | - | 25 |
| TC | - | - | 2 | - | 2 |
| Total | 38 | 15 | 2 | 1 | 56 |

Sourced Material from Tikal/Yaxha by Time Period

Table 2.30

Some material was identified by Ford *et al.* as belonging to mixed Early/Late Classic deposits. These pieces were, for the purposes of this study, considered with the Late Classic material when analysed.

Obsidian from the Tikal/Yaxha project is weighted toward the Late Classic Period with 66.1 per cent of the collection being dated to this period, and an additional 3.5 per cent coming from the Terminal Classic Period. Material from the Early Classic accounted for only 7.2 per cent of the sample, although this may be partly the result of the material that Ford *et al.* could not securely date and classified merely as "Classic" Period (Ford *et al.* 1997). The Formative Period material composed the second largest category with 23.2 per cent of the collection.

Sources utilised in these deposits are consistent with patterns found elsewhere. San Martin Jilotepeque material forms the majority of the Formative collection (76.9 per cent), with evidence of utilisation of El Chayal material (23.1 per cent). Evidence from the Early Classic is suspect, as there are only four pieces, with equal amounts of El Chayal and Ixtepeque material (25 per cent each) and a larger quantity of San Martin Jilotepeque material (50 per cent) than is usual. The Late Classic sample demonstrates a typical pattern of high reliance on El Chayal material (87.2 per cent), with both San Martin Jilotepeque and Ixtepeque material (7.7 per cent and 5.1 *per cent* respectively) represented in lesser degrees. The reliance on Ixtepeque material tends to increase toward the Post-Classic; this may be what is demonstrated in the Tikal/Yaxha collection, as the obsidian found in the Terminal Classic deposits was Ixtepeque. However, as with the Early Classic collection, this small sample must be considered with care. Analysing material from more than one site assists in providing a large enough database from which reliable conjectures regarding the reliance upon sources may be created.

### SUMMARY OF OBSIDIAN DISTRIBUTION

The above discussions briefly outline the sites chosen for this study and their appropriate geographical categories. Obsidian, as is clear from the above studies, is a resource acquired by virtually every type of site, large and small, in almost every period. The ubiquitousness of this material is what makes it ideal for intra-site distributional studies. It is clear from the above site summaries that the pattern commonly noted in obsidian source utilisation is present in the material selected for this study.

Middle Formative Period obsidian exchange was dominated by material from the San Martin Jilotepeque source. Reliance shifted away from this source in the Late Formative period with the rise of El Chayal material. During the Early and Late Classic Periods El Chayal remained the favoured material with significantly higher proportions of the collections. This pattern is present in each of the areas examined. During the Terminal Classic, although the presence of El Chayal material remains high, material from Ixtepeque increases in prominence.

Identifying the pattern of obsidian source utilisation is a commonly practised exercise in obsidian studies. The question that arises from this pattern, and one that is rarely addressed, is *why does this pattern appear?* It is my opinion that this question can best be answered by examining the archaeological and social contexts of the material. It is only through this type of analysis that we can hope to acquire the necessary information on *who* was using obsidian and *for what purposes.* Without knowing the *who* and *what* of obsidian utilisation, understanding the *why* of obsidian reliance and exchange will always be beyond our reach, and any theories that are produced regarding this issue will remain theories.

The following section presents the results of investigations into the *who* and *what* of obsidian utilisation. These studies are conducted on the material holistically, as well as by geographical category, and include discussions on alternative classificatory and analytical processes for obsidian studie

## CHAPTER 3
## DATA PARADIGMS AND TESTS

### INTRODUCTION

As noted at the close of the previous section, the purpose of this work is to investigate possible correlations between reliance upon various obsidian sources and distribution of artefacts among different contexts. To this end, the material from the sites outlined in the previous section will be subjected to a variety of computations revolving around the variables source, time, quality, and functional, archaeological, and social contexts. Each computation will analyse the material holistically, then dissect the model into the geographical regions outlined in Chapter 2 Section 2. It is predicted that this process will reveal general patterns and variations that might otherwise be subsumed in a larger analysis.

The first computation will present data demonstrating variations in source utilisation through time. This process is designed to demonstrate the broad nature of the patterns of obsidian consumption. Following the identification of this pattern, the material will be analysed by including the variable 'context' into the paradigm. It is expected that through the addition of this variable, the underlying nature of obsidian source reliance may be understood. The third analysis will investigate possible correlates between different grades, or qualities, of obsidian and their distribution between the contexts utilised in the second process. The final analysis will involved a discussion of the relevance of cutting edge to mass ratios (CE:M) as a means of understanding the consumption pattern of obsidian at an intra-site level.

### SECTION 1
### SOURCE AND TIME DISTRIBUTION PATTERN

While the obsidian source consumption pattern analysed holistically at the end of the previous section (Table 3.1) displays the predicted model noted in other monographs (Asaro and Michel 1981; Asaro *et al.* 1987; Cobean 1991; Dreiss 1986; Dreiss and Brown 1989; Moholy-Nagy 1975, 1984; Moholy-Nagy and Nelson 1990; Nelson 1985; Nelson and Howard 1986; Nelson *et al.* 1983), it is important to examine this material regionally to note any divergences that may cause the paradigm to vary.

#### FORMATIVE PERIODS

While the shift in source utilisation from San Martin Jilotepeque to El Chayal in the Late Formative Period is generally recognised, the dramatic nature of this shift may be questioned. Initial impressions of the material (Table 3.1) suggest that the shift between these two sources during the Middle Formative (MF) and Late Formative (LF) Periods was quite rapid with San Martin Jilotepeque material decreasing from 74.7 per cent in the Middle Formative to 14.8 per cent. During this same time, El Chayal material increases from 16.0 per cent to 71.9 per cent. These figures represent a decline in consumption of San Martin Jilotepeque material by roughly 60 per cent and increase in reliance on El Chayal material by close to 70 per cent.

However, when the obsidian is divided into its regional components, a different picture emerges (Tables 3.2 to 3.5). Shifts from San Martin Jilotepeque material to El Chayal material for the same time frames are more gradual, with El Chayal material among the Inland Sites rising only 35 per cent between the Middle and Late Formative Periods and San Martin Jilotepeque material decreasing during the same period by only 5 per cent. The gradual nature of the shift is also apparent in the material from the Central Peten. In this area, consumption of El Chayal obsidian increases by only 25.5 per cent, while San Martin Jilotepeque material decreases by only 31.7 per cent. These increases and decreases create a situation in the Late Formative Period where consumption of the San Martin Jilotepeque and El Chayal obsidian is almost equal at 43.3 and 42.7 per cent of the collection respectively.

The two abnormalities noted in the patterns of consumption are in the Coastal region and among the River Sites. No San Martin Jilotepeque material was identified in the former area, and no material from the Middle Formative Period was available for analysis. In the latter area, the increase between the Middle and Late Formative Periods is quite dramatic.

| | ELC (%) | SMJ (%) | IXT (%) | MX (%) | UNK (%) | Total (%) |
|---|---|---|---|---|---|---|
| MF | 12 (16.0) | 56 (74.7) | 5 (6.7) | - | 2 (2.6) | 75 (3.9) |
| LF | 511 (71.9) | 105 (14.8) | 72 (10.1) | 2 (0.3) | 21 (2.9) | 711 (37.6) |
| EC | 307 (83.0) | 17 (4.6) | 16 (4.3) | 29 (7.8) | 4 (0.3) | 373 (19.7) |
| LC | 424 (73.2) | 40 (6.9) | 93 (16.1) | 15 (2.6) | 7 (1.2) | 579 (30.6) |
| TC | 79 (51.0) | 6 (3.9) | 63 (40.6) | 3 (1.9) | 4 (2.6) | 155 (8.2) |
| Total | 1333 | 224 | 249 | 49 | 38 | 1893 (100.0) |

Total Obsidian Available by Source and Time Period
Table 3.1

| | ELC (%) | SMJ (%) | IXT (%) | MX (%) | UNK %) | Total (%) |
|---|---|---|---|---|---|---|
| MF | - | 7 (100) | - | - | - | 7 (0.7) |
| LF | 416 (80.6) | 31 (6.0) | 48 (9.3) | - | 21 (4.1) | 516 (49.5) |
| EC | 162 (92.6) | 3 (1.7) | 9 (5.1) | - | 1 (0.6) | 175 (16.8) |
| LC | 129 (58.4) | 12 (5.4) | 79 (35.7) | 1 (0.5) | - | 221 (21.2) |
| TC | 62 (50.0) | 5 (4.0) | 57 (46.0) | - | - | 124 (11.9) |
| Total | 769 | 58 | 193 | 1 | 22 | 1043 (100.1) |

Obsidian from River-Based Sites by Source and Time Period

Table 3.2

| | ELC (%) | SMJ (%) | IXT (%) | MX (%) | Total (%) |
|---|---|---|---|---|---|
| MF | 1 (25) | 1 (25) | 2 (50) | - | 4 (1.7) |
| LF | 9 (60) | 3 (20) | 3 (20) | - | 15 (6.5) |
| EC | 77 (82.8) | 1 (1.1) | 2 (2.2) | 13 (14.0) | 93 (40.4) |
| LC | 113 (96.6) | - | 4 (3.4) | - | 117 (50.9) |
| TC | 1 (100) | - | - | - | 1 (0.4) |
| Total | 201 | 5 | 11 | 13 | 230 (99.9) |

Obsidian from Inland Sites by Source and Time Period

Table 3.3

| | ELC (%) | SMJ (%) | IXT (%) | MX (%) | UNK (%) | Total (%) |
|---|---|---|---|---|---|---|
| MF | 11 (17.2) | 48 (75.0) | 3 (4.7) | - | 2 (3.1) | 64 (11.3) |
| LF | 70 (42.7) | 71 (43.3) | 21 (12.8) | 2 (1.2) | - | 164 (29.0) |
| EC | 54 (62.1) | 13 (14.9) | 1 (1.1) | 16 (18.4) | 3 (3.4) | 87 (15.4) |
| LC | 165 (74.7) | 28 (12.7) | 7 (3.2) | 14 (6.3) | 7 (3.2) | 221 (39.1) |
| TC | 15 (51.7) | 1 (3.4) | 6 (20.7) | 3 (10.3) | 4 (13.8) | 29 (5.1) |
| Total | 315 | 161 | 38 | 35 | 16 | 565 (99.9) |

Obsidian from Central Peten Sites by Source and Time Period

Table 3.4

| | ELC (%) | IXT (%) | Total (%) |
|---|---|---|---|
| MF | - | - | - |
| LF | 16 (100) | - | 16 (29.1) |
| EC | 14 (77.8) | 4 (22.2) | 18 (32.7) |
| LC | 17 (85) | 3 (15) | 20 (36.4) |
| TC | 1 (100) | - | 1 (1.8) |
| Total | 48 | 7 | 55 (100) |

Obsidian from Coastal Sites by Source and Time Period

Table 3.5

Among River sites, consumption of El Chayal material increases from none in the Middle Formative to 416 pieces, 80.6 per cent in the Late Formative collection. Although consumption of San Martin Jilotepeque material decreases from 100 per cent to 6 per cent during this same time, these quantities are deceptive as the San Martin Jilotepeque material actually increases from 7 to 31 pieces when considered in terms of raw counts. This increase in material is as suggestive of a limited Middle Formative database as it is of a general increase in Late Formative obsidian consumption. However, the higher than expected quantities of El Chayal in the Late Formative deposits at River sites is a pattern worthy of further, contextual investigation, which will be discussed in the section on contexts below.

A point of interest may be found in the surprising quantities of Ixtepeque material in the Late Formative Period. This material accounts for 10.1 per cent of the total Late Formative collection, only slightly lower than San Martin Jilotepeque material, and throws into question the argument that this source was not a significant production area until the Late Classic. The majority of this material, 48 of the 72 pieces, is associated with River Sites, where it comprises 9.3 per cent of the collection. As Hammond (1972) postulates that Ixtepeque material was traded via coastal and river routes, the presence of this material at river sites should not be surprising. However, a paucity of material from Coastal sites makes correlating deposits along this trade route with the hypothesised synchronous Coastal route problematic. This material is also present in quantities of note at Central Peten sites, where it represents 12.8 per cent of the collection. A limited amount of Mexican obsidian (1.2 per cent, n=2) also appears at sites in the Central Peten during this period.

### EARLY CLASSIC

During the Early Classic (EC) period, the distribution of obsidian continues to shift in favour of El Chayal, increasing from 71.9 per cent in the Formative Period to 83.0 per cent of the Early Classic collection. This shift is significantly more gradual, at only 11.1 per cent, than the previous increase of 55.9 per cent between the two Formative periods. Among River sites, the area which previously had demonstrated the greatest increase, El Chayal material rose by only 12 percent. In the Inland and Central Peten areas, the increase were a more even 22.8 and 19.4 per cent, respectively. Obsidian from Mexican sources increases in this period, rising from 0.3 per cent (n=2) in the Late Formative to 7.8 per cent (n=29). This material is split almost equally between Central Peten and Inland sites, where it represents 18.4 per cent (n=16) and 14 per cent (n=13) of the respective collections. Both San Martin Jilotepeque and Ixtepeque material decline in all regions during this period.

### LATE CLASSIC

Although El Chayal material continues to predominate in the Late Classic (LC) Period, the amount of the total of this material declines slightly to 73.2 per cent. Mexican material also declines overall from 7.8 per cent to 2.6 per cent. Material from San Martin Jilotepeque and Ixtepeque both increase, the San Martin material to 6.9 per cent and the Ixtepeque material to 16.1 per cent of the total collection.

Increase in consumption of Ixtepeque material is highest in the River region, where it accounts for 35.7 per cent of the collection, up from 5.1 per cent in the previous period. As noted in the discussion of Formative deposits, the presence of this material along river routes should not be considered unusual. Small quantities of this material were found in sites along the coast during this period, where it accounted for 15 per cent of the material from that region. However, it should be noted that in general the collection of material from the coast is too small to make conclusive regional statements. Although Ixtepeque material increases in the Central Peten and Inland regions, it represents significantly smaller portions of these collections, 3.2 per cent and 3.4 per cent respectively.

Reliance on San Martin Jilotepeque material increases only in the River region, rising from 1.7 per cent of the Early Classic collection to 5.4 per cent of the Late Classic collection. In the Central Peten area, it decreases by 2.2 per cent, and is absent in both the Inland and Coastal regions. In the latter area, its absence is not unusual, as no San Martin Jilotepeque material was reported for any of the periods.

Mexican material disappears in the Inland region, where it had previously accounted for 14 per cent of the collection. Although still present in the Central Peten area, Mexican obsidian declines to 6.3 per cent of the collection, down from 18.4 per cent. A single piece was reported from the River region, accounting for 0.5 per cent of collection.

### TERMINAL CLASSIC

El Chayal material continues to decline in prominence in during this period, dropping to 51 per cent of the total collection. Materials from San Martin Jilotepeque and Mexico also decline, accounting for 3.9 per cent and 1.9 per cent of the total Terminal Classic collection, respectively. The only source for which consumption increases is Ixtepeque, which rises from 16.1 per cent to 40.6 per cent of the collection. This increase corresponds with previously predicted patterns that suggest a general shift away from El Chayal material toward Ixtepeque in the Late and Post Classic period.

While all regions reported Terminal Classic deposits, the Inland and Coastal regions recorded only a single artefact apiece, making these collections too small for meaningful discussion. In the River and Central Peten areas, the quantities of El Chayal material roughly match the pattern predicted in the holistic model (50.0 per cent and 51.7 per cent respectively).

Quantities of Ixtepeque material increase in both the River and Central Peten regions during this period. Although the quantity of Ixtepeque material in the River area is more than double that of the Central Peten (46 per cent compared to 20.4 per cent), it is this latter region that exhibited the largest increase. Consumption of Ixtepeque material increased by 17.2 per cent in the Central Peten area, compared to an increase of 10.3 per cent in the River region.

#### SUMMARY

While when analysed holistically the pattern of obsidian consumption appears to fit the previously document paradigm for obsidian source distribution, the dissection of the material into regional levels exhibits a secondary and somewhat contradictory pattern.

The most noticeable variation in this pattern is the absence of San Martin Jilotepeque material from the Coastal region. While this absence may be explained by Hammond's assertion (1972) that San Martin Jilotepeque material travelled via the Chixoy/Passion River and was therefore traded primarily in the interior of the Yucatan, this pattern may also be explained, perhaps more prosaically, by a paucity of samples. However, regardless of the exact rationale behind the absence of the material, it is important to note that patterns previously outlined for Maya obsidian consumption do not apply to this region.

Another element that does not fit into the generally prescribed pattern of shifting consumption are the low quantities of Ixtepeque material recovered from the Inland region. The 'classic' pattern predicted by the holistic model actually appears only in the Central Peten and River region. The latter region is of particular interest as the shifts in source utilisation appear more elastic than those in other regions. The following section will attempt to clarify these patterns further with the addition of contextual data in the analytical matrix.

## SECTION 2
## SOURCE, CONTEXT, AND TIME MODEL

The previous section discussed the distribution of artefacts from various obsidian sources through time, both holistically and in terms of geographical regions outlined in Chapter 2, Section 2. The purpose of this section is to investigate the distribution patterns outlined previously to determine possible reasons why these shifts occur. Two questions were posed in regards to the data that are thought best to elucidate why obsidian source consumption assumes the pattern it does. These questions are 1) do some sources appear more frequently or exclusively in some contexts, and, 2) does this pattern change through time?
Material was viewed in terms of its functional, archaeological, and social contexts, which are outlined in detail in Chapter 2, Section 1. The purpose of these categories is to determine if the pattern of consumption for different sources of obsidian is linked to specific functions or to particular strata in society, and if these patterns vary temporally and/or geographically.

### DISCUSSION OF FUNCTIONAL CONTEXTS

Broad in scope, these contexts are chosen to illuminate patterns in obsidian consumption on a general level. Modelled after Schiffer's ideas of archaeological and systemic transfers of material, four 'functional' contexts were identified, with an additional category for material of indeterminate purpose. Functional contexts included 1) ritual, 2) waste, 3) domestic, and 4) burial. Although some burials may incorporate ritual aspects, the questionable nature of artefacts included in a grave, personal possession versus *prestation* (Mauss 1994), warranted a separate category for this material. The fifth category was simply dubbed 'indeterminate', an indication perhaps of its modern function if not its original one. Material was examined both holistically and regionally to determine if, and possibly why, variations may occur.

#### MIDDLE FORMATIVE DEPOSITS

Only two regions yielded material dated to this period: the Central Peten and the Inland zone. All obsidian material from both regions was discovered in 'waste' deposits. The majority of this material was from the San Martin Jilotepeque source (74.7 per cent). Other artefacts recovered included objects produced from El Chayal and Ixtepeque material (16 per cent and 6.7 per cent respectively), along with some material that was unidentifiable (2.6 per cent).

Middle Formative deposits were reported in two locations: the Inland and Peten areas. The Inland area yielded only four pieces of obsidian, all from waste contexts. The majority of the material was recovered from the Central Peten Lakes, a project that tested 355 structures around six lakes. The report suggests that all the material recovered, including the untested obsidian, was discovered in waste deposits (mixed construction fill and midden). Although suggestive of a behaviour pattern that does not conserve or covet obsidian as a ritual good, but rather considers the material a more utilitarian good. The limited sample merits caution, however, as, outlined in Chapter 1 Section2, single deposits may not represent a site-wide pattern but rather a particular event or social or political issue. However, the presence of large quantities of San Martin Jilotepeque material in waste contexts is the first piece in a pattern that is worthy of note.

### LATE FORMATIVE DEPOSITS

Unlike the previous period, material from the Late Formative was dispersed throughout the five categories. As noted previously, this period saw a marked increase in consumption of El Chayal material. Analysis of the distribution pattern in this period shows that the largest majority of the El Chayal material, 326 of the 445 pieces (73.3 per cent of the El Chayal material and 58 per cent of the collection), was utilised in ritual contexts. The material was distributed between three different caches, placed in three different structures, in two distinct areas of the Blue Creek Site: two in the core area and one off the escarpment in the non-elite settlement zone. While the large presence of El Chayal material in ritual contexts is interesting, it becomes more significant when one realises that this source is the *only* material recovered in this context. It is also the *only* material recovered from burial contexts.

Other sources utilised during this period include San Martin Jilotepeque, Ixtepeque and material described as "other". This latter category may include San Martin Jilotepeque material as the material from Cuello, which composes the entirety of the 'other' collection here, was not tested for San Martin Jilotepeque material (Hammond 1991b). Material from these other categories are concentrated primarily in waste deposits, with some material recovered from domestic contexts and one piece from an indeterminate deposit (Table 3.6). This pattern of exclusive use of El Chayal in ritual contexts is clearest in the material from Blue Creek, as this is the only site that reports obsidian recovered from ritual activities for this period. However, we do know that ritual activity was on the increase during this period (Freidel and Schele 1988a, 1988b). Other regions note the predominance of materials in waste deposits, with El Chayal material being the sole source utilised as mortuary goods.

Several things may be suggested by this consumption pattern. The most apparent phenomenon is an increase in ritual activity. This increase is most notable at Blue Creek, but similar increases may be implied for other areas. The statement regarding ritual activity may be followed by an equally obvious statement noting this activity demonstrated a preference for El Chayal material. The dominance of El Chayal material in ritual and, to a lesser extent, burial contexts coupled with the concentration of other materials, specifically San Martin Jilotepeque material, in waste and domestic contexts is the second key in understanding my suggested paradigm.

### EARLY CLASSIC DEPOSITS

Obsidian distribution during this period mimicked the pattern previously identified. El Chayal objects continued to dominate the material utilised and, as in the previous period, was distributed throughout the various contexts, with the majority of material (70.3 per cent, n=189) concentrated in ritual deposits. Ixtepeque material was also encountered in all identifiable contexts; however, the majority of material was found in waste deposits (71.4 per cent, n=10). Objects manufactured from material from central Mexican sources are introduced into the region during this period. Despite arguments that this material, due to it remote and "exotic" origin should be more highly coveted than 'local' objects, the majority of the material (69.6 per cent of Mexican material, n=16) was found in waste contexts. Only a limited amount was found in burials or ritual deposits (13 per cent each, n=3).

Those regions that include obsidian from ritual activities include the Central Peten, Inland, and River areas. In the Central Peten, the only material noted in ritual contexts was El Chayal in origin. Within the Inland area, both El Chayal and Mexican obsidian were used in ritual deposits: however, the majority of the objects were El Chayal in origin (94.4 per cent, n=51). In the River area, 137 objects were recovered from ritual contexts. Of these, 135 (98.5 per cent) were manufactured from El Chayal obsidian. The other two items were manufactured from San Martin Jilotepeque and Ixtepeque material.

With the exception of a single piece, all objects manufactured from San Martin Jilotepeque material were found in waste contexts. This equates to 90 per cent of the San Martin Jilotepeque collection from this period. The single anomalous piece was discovered in a large ritual deposit at Blue Creek. This continues the trend noted for the previous periods, where San Martin Jilotepeque material is primarily in utilitarian or waste contexts.

### LATE CLASSIC DEPOSITS

Distribution of objects from the various sources during this period continues in the same pattern outlined with only slight variations. Although the collection remains dominated by material from the El Chayal source, distribution of these objects shifts away from ritual context, with more items being found in waste and domestic contexts. Ritual deposits consume only 48 per cent of the El Chayal material of this period, a substantial drop from the 70.3 per cent of the previous period. An almost equal quantity of El Chayal material (42.3 per cent) was recovered from waste contexts, and domestic consumption doubled from the previous period, rising from 2.6 per cent to 5.3 per cent of the El Chayal sample. San Martin Jilotepeque material is still concentrated in waste contexts which, as in the previous period, utilised 90 per cent of this material. The other 10 per cent was distributed evenly between domestic and indeterminate contexts.

Objects manufactured from Ixtepeque material were primarily recovered from waste deposits, with 55.6 per cent of items from this source coming from this context. A large portion of Ixtepeque material (28.4 per cent) was recovered from indeterminate deposits, while only a single piece (1.2 per cent) was recovered from ritual contexts. Mexican materials were also concentrated in waste deposits during this period.

### Terminal Classic Deposits

Information from this period is more sketchy than those previous as only a few of the site reports utilised in this study made use of this period. However, it is of interest to note that the decline in ritual activity noted for the Late Classic appears to continue, with no such deposits being recorded for this period among the sites used in this study. Very limited amounts of San Martin Jilotepeque material were attributed to this period and these pieces were reported in either waste or indeterminate contexts. Both El Chayal and Ixtepeque sources were heavily utilised, accounting for 54.6 per cent and 41.9 per cent of the Late Classic collection respectively. Objects manufactured from these sources were found in waste, burial, and indeterminate contexts. The large quantities of items recovered in indeterminate contexts may be attributed to their proximity to or location in humus layers. Of note is that only objects from these two sources were included in burial contexts. Items from the other two sources were in waste or indeterminate contexts.

### Summary of Functional Contexts

Analysis of the material by functional contexts revealed two complementary patterns of source utilisation. The first and most noticeable of these is the preferential utilisation of El Chayal material in Ritual and Burial contexts. During the Late Formative, ritual and burial contexts consisted entirely of El Chayal material. In the Early Classic, 194 pieces of obsidian were recovered from ritual contexts, of which 189 pieces, or 97.4 per cent, were El Chayal in origin. Single pieces of Ixtepeque and San Martin Jilotepeque obsidian, amounting to 0.5 per cent of the ritual collection each, were also recovered, as were three items of Mexican. This piece of San Martin Jilotepeque material is the only item manufactured from this source that was found in a ritual context in *any* period. The dominance of El Chayal material in ritual contexts continues in the Late Classic, where 154 pieces (98.7 per cent) of the 156 pieces recovered in these deposits are from this source. Single pieces of Ixtepeque and Mexican obsidian, accounting for 0.6 per cent each, were also recovered from Late Classic
ritual contexts. Although El Chayal material is the dominate source utilised in the Late Formative, Early Classic and Late Classic periods (71.9 per cent, 83.0 per cent, and 73.2 per cent respectively), the proportions of this material recovered in ritual contexts (100 per cent, 97.4 per cent, and 98.7 per cent respectively) greatly exceed the general distribution pattern. This suggests that El Chayal material was the material of choice for ritual contexts. That this material is also found in other context may be seen as a "trickle down effect", where highly coveted or valued material is sought by all levels of society, but lower status or poor material is not considered for higher status activities.

This brings us to the second pattern, the lack of San Martin Jilotepeque material in Ritual or Burial contexts. With the exception of a single item manufacture from San Martin Jilotepeque material, all items from this source are concentrated in waste or utilitarian deposits. This material appears only the once in a ritual context and never in burial contexts.

An interesting point of note is the paucity of Mexican material in Ritual or Burial contexts. In both the Early Classic and Late Classic Periods (the only two in which material of Mexican origin was recovered), the majority of objects manufactured of Mexican obsidian are in waste contexts, 69.6 per cent and 80.0 per cent of the respective collections. Moholy-Nagy notes a similar pattern at Tikal, and surmises that "the apparent scarcity of Mexican obsidian in burials and caches indicates that the elite did not consider it either a prestige good or of ritual value" (Moholy-Nagy 1999:310). Alternatively, the lack of Mexican obsidian in ritual or burial deposits may be less an indication of the value of the material than of the role it fulfilled in Maya society. As a 'foreign' good, it might be of greater value moving within systemic contexts than isolated and static in a cache.

| Context | Middle Formative | | | | Late Formative | | | | Early Classic | | | | | Late Classic | | | | | Terminal Classic | | | |
|---|---|---|---|---|---|---|---|---|---|---|---|---|---|---|---|---|---|---|---|---|---|---|
| | EC | SM | IX | OT | EC | SM | IX | OT | EC | SM | IX | MX | OT | EC | SM | IX | MX | OT | EC | SM | IX | OT |
| Ritual | - | - | - | - | 326 | - | - | - | 189 | 1 | 1 | 3 | - | 154 | - | 1 | 1 | - | - | - | - | - |
| Waste | 12 | 48 | 5 | 2 | 99 | 62 | 35 | 13 | 65 | 9 | 10 | 16 | 2 | 136 | 18 | 45 | 4 | 3 | 27 | 1 | 30 | 3 |
| Domestic | - | - | - | - | 14 | 4 | 2 | - | 7 | - | 2 | 1 | - | 17 | 1 | 12 | - | - | - | - | - | - |
| Burial | - | - | - | - | 3 | - | - | - | 8 | - | 1 | 3 | 2 | - | - | - | - | - | 21 | - | 12 | - |
| Ind | - | - | - | - | 3 | 1 | - | - | - | - | - | - | - | 14 | 1 | 23 | - | - | 29 | 1 | 17 | - |
| *Total 1* | *12* | *48* | *5* | *2* | *445* | *67* | *37* | *13* | *269* | *10* | *14* | *23* | *4* | *321* | *20* | *81* | *5* | *3* | *77* | *2* | *59* | *3* |
| *Total 2* | *67* | | | | *562* | | | | *320* | | | | | *430* | | | | | *141* | | | |

Quantity of Material from All Sites Categorised by Functional Context, Source, and Time Period

Table 3.6

| Context | Middle Formative | | | | Late Formative | | | | Early Classic | | | | | Late Classic | | | | | Terminal Classic | | | |
|---|---|---|---|---|---|---|---|---|---|---|---|---|---|---|---|---|---|---|---|---|---|---|
| | EC | SM | IX | OT | EC | SM | IX | OT | EC | SM | IX | MX | OT | EC | SM | IX | MX | OT | EC | SM | IX | OT |
| Ritual | - | - | - | - | 58.0 | - | - | - | 59.2 | 0.3 | 0.3 | 0.9 | - | 35.8 | - | 0.2 | 0.2 | - | - | - | - | - |
| Waste | 17.9 | 71.6 | 7.5 | 3.0 | 17.6 | 11.0 | 6.2 | 2.3 | 20.4 | 2.8 | 3.1 | 5.0 | 0.6 | 31.6 | 4.2 | 10.5 | 0.9 | 0.7 | 19.1 | 0.7 | 21.3 | 2.1 |
| Domestic | - | - | - | - | 2.5 | 0.7 | 0.4 | - | 2.2 | - | 0.6 | 0.3 | - | 4.0 | 0.2 | 2.8 | - | - | - | - | - | - |
| Burial | - | - | - | - | 0.5 | - | - | - | 2.5 | - | 0.3 | 0.9 | 0.6 | - | - | - | - | - | 14.9 | - | 8.5 | - |
| Ind | - | - | - | - | 0.5 | 0.2 | - | - | - | - | - | - | - | 3.3 | 0.2 | 5.3 | - | - | 20.6 | 0.7 | 12.1 | - |
| *Total 1* | *17.9* | *71.6* | *7.5* | *3.0* | *79.1* | *11.9* | *6.6* | 2.3 | *84.1* | *3.1* | *4.3* | *7.1* | *1.2* | *74.7* | *4.6* | *18.8* | *1.1* | *0.7* | *54.6* | *1.4* | *41.9* | *2.1* |
| *Total 2* | *100.0* | | | | *99.9* | | | | *99.8* | | | | | *99.9* | | | | | *100.0* | | | |

Percentages of Material from All Sites Categorised by Functional Context, Source, and Time Period

Table 3.7

| Context | Middle Formative | | | | Late Formative | | | | Early Classic | | | | | Late Classic | | | | | Terminal Classic | | | |
|---|---|---|---|---|---|---|---|---|---|---|---|---|---|---|---|---|---|---|---|---|---|---|
| | EC | SM | IX | OT | EC | SM | IX | OT | EC | SM | IX | MX | OT | EC | SM | IX | MX | OT | EC | SM | IX | OT |
| Ritual | - | - | - | - | - | - | - | - | 3 | - | - | - | - | - | - | - | - | - | - | - | - | - |
| Waste | 11 | 47 | 3 | 2 | 27 | 38 | 4 | - | 20 | 8 | - | 14 | 1 | 75 | 16 | 4 | 4 | 3 | 13 | 1 | 2 | 3 |
| Domestic | - | - | - | - | 7 | 3 | - | - | - | - | - | 1 | - | 3 | - | - | - | - | - | - | - | - |
| Burial | - | - | - | - | 1 | - | - | - | 2 | - | - | 3 | 2 | - | - | - | - | - | - | - | - | - |
| Ind | - | - | - | - | - | - | - | - | - | - | - | - | - | - | - | - | - | - | - | - | - | - |
| *Total 1* | *11* | *47* | *3* | *2* | *35* | *41* | *4* | - | *25* | *8* | - | *18* | *3* | *84* | *16* | *4* | *4* | *3* | *13* | *1* | *2* | *3* |
| *Total 2* | *63* | | | | *80* | | | | *54* | | | | | *111* | | | | | *19* | | | |

Quantity of Material from Peten Sites Categorised by Functional Context, Source, and Time Period

Table 3.8

| Context | Middle Formative | | | | Late Formative | | | | Early Classic | | | | | Late Classic | | | | | Terminal Classic | | | |
|---|---|---|---|---|---|---|---|---|---|---|---|---|---|---|---|---|---|---|---|---|---|---|
| | EC | SM | IX | OT | EC | SM | IX | OT | EC | SM | IX | MX | OT | EC | SM | IX | MX | OT | EC | SM | IX | OT |
| Ritual | - | - | - | - | - | - | - | - | 5.5 | - | - | - | - | 5.4 | - | - | - | - | - | - | - | - |
| Waste | 17.5 | 74.5 | 4.8 | 3.2 | 33.7 | 47.5 | 5.0 | - | 37.0 | 14.8 | - | 25.9 | 1.8 | 67.6 | 14.4 | 3.6 | 3.6 | 2.7 | 68.4 | 5.3 | 10.5 | 15.8 |
| Domestic | - | - | - | - | 8.7 | 3.7 | - | - | - | - | - | 1.8 | - | 2.7 | - | - | - | - | - | - | - | - |
| Burial | - | - | - | - | 1.2 | - | - | - | 3.7 | - | - | 5.5 | 3.7 | - | - | - | - | - | - | - | - | - |
| Ind | - | - | - | - | - | - | - | - | - | - | - | - | - | - | - | - | - | - | - | - | - | - |
| *Total 1* | *17.5* | *74.6* | *4.8* | *3.2* | *43.6* | *51.2* | *5.0* | - | *46.2* | *14.8* | - | *33.2* | *5.5* | *75.7* | *14.4* | *3.6* | *3.6* | *2.7* | *68.4* | *5.3* | *10.5* | *15.8* |
| *Total 2* | *100.1* | | | | *99.8* | | | | *99.7* | | | | | *100.0* | | | | | *100.0* | | | |

Percentages of Material from Peten Categorised by Functional Context, Source, and Time Period

Table 3.9

| Context | Middle Formative | | | | Late Formative | | | | Early Classic | | | | | Late Classic | | | | |
|---|---|---|---|---|---|---|---|---|---|---|---|---|---|---|---|---|---|---|
| | EC | SM | IX | OT | EC | SM | IX | OT | EC | SM | IX | MX | OT | EC | SM | IX | MX | OT |
| Ritual | - | - | - | - | - | - | - | - | 51 | - | - | 3 | - | 86 | - | - | - | - |
| Waste | 1 | 1 | 2 | - | 1 | - | 1 | - | 18 | 2 | 2 | - | - | 1 | - | - | - | - |
| Domestic | - | - | - | - | - | - | - | - | - | - | - | - | - | - | - | - | - | - |
| Burial | - | - | - | - | - | - | - | - | - | - | - | - | - | - | - | - | - | - |
| Ind | - | - | - | - | - | - | - | - | - | - | - | - | - | - | - | - | - | - |
| *Total 1* | *1* | *1* | *2* | - | *1* | - | *1* | - | *69* | *2* | *2* | *3* | - | *87* | - | - | - | - |
| *Total 2* | *4* | | | | *2* | | | | *76* | | | | | *87* | | | | |

Quantity of Material from Inland Sites Categorised by Functional Context, Source, and Time Period

Table 3.10

| Context | Middle Formative | | | | Late Formative | | | | Early Classic | | | | | Late Classic | | | | |
|---|---|---|---|---|---|---|---|---|---|---|---|---|---|---|---|---|---|---|
| | EC | SM | IX | OT | EC | SM | IX | OT | EC | SM | IX | MX | OT | EC | SM | IX | MX | OT |
| Ritual | - | - | - | - | - | - | - | - | 67.1 | - | - | 3.9 | - | 98.9 | - | - | - | - |
| Waste | 25.0 | 25.0 | 50.0 | - | 50.0 | - | 50.0 | - | 23.7 | 2.6 | 2.6 | - | - | 1.1 | - | - | - | - |
| Domestic | - | - | - | - | - | - | - | - | - | - | - | - | - | - | - | - | - | - |
| Burial | - | - | - | - | - | - | - | - | - | - | - | - | - | - | - | - | - | - |
| Ind | - | - | - | - | - | - | - | - | - | - | - | - | - | - | - | - | - | - |
| *Total 1* | *25.0* | *25.0* | *50.0* | - | *50.0* | - | *50.0* | - | *90.8* | *2.6* | *2.6* | *3.9* | - | - | - | - | - | - |
| *Total 2* | *100.0* | | | | *100.0* | | | | *99.9* | | | | | *100.0* | | | | |

Percentages of Material from Inland Sites Categorised by Functional Context, Source, and Time Period

Table 3.11

| Context | Late Formative | | | | Early Classic | | | | | Late Classic | | | | | Terminal Classic | | | |
|---|---|---|---|---|---|---|---|---|---|---|---|---|---|---|---|---|---|---|
| | EC | SM | IX | OT | EC | SM | IX | MX | OT | EC | SM | IX | MX | OT | EC | SM | IX | OT |
| Ritual | - | - | - | - | - | - | - | - | - | - | - | - | - | - | 1 | - | - | - |
| Waste | 16 | - | - | - | 9 | - | 1 | - | - | 17 | - | 3 | - | - | - | - | - | - |
| Domestic | - | - | - | - | - | - | - | - | - | - | - | - | - | - | - | - | - | - |
| Burial | - | - | - | - | 4 | - | 1 | - | - | - | - | - | - | - | - | - | - | - |
| Ind | - | - | - | - | - | - | - | - | - | - | - | - | - | - | - | - | - | - |
| *Total 1* | *16* | - | - | - | *13* | - | *2* | - | - | *17* | - | *3* | - | - | *1* | - | - | - |
| *Total 2* | *16* | | | | *15* | | | | | *20* | | | | | *1* | | | |

Quantity of Material from Coastal Sites Categorised by Functional Context, Source, and Time Period

Table 3.12

| Context | Late Formative | | | | Early Classic | | | | | Late Classic | | | | | Terminal Classic | | | |
|---|---|---|---|---|---|---|---|---|---|---|---|---|---|---|---|---|---|---|
| | EC | SM | IX | OT | EC | SM | IX | MX | OT | EC | SM | IX | MX | OT | EC | SM | IX | OT |
| Ritual | - | - | - | - | - | - | - | - | - | - | - | - | - | - | 100.0 | - | - | - |
| Waste | 100.0 | - | - | - | 60.0 | - | 6.7 | - | - | 85.0 | - | 15.0 | - | - | - | - | - | - |
| Domestic | - | - | - | - | - | - | - | - | - | - | - | - | - | - | - | - | - | - |
| Burial | - | - | - | - | 26.7 | - | 6.7 | - | - | - | - | - | - | - | - | - | - | - |
| Ind | - | - | - | - | - | - | - | - | - | - | - | - | - | - | - | - | - | - |
| *Total 1* | *100.0* | - | - | - | *86.7* | - | *13.4* | - | - | *85.0* | - | *15.0* | - | - | *100.0* | - | - | - |
| *Total 2* | *100.0* | | | | *100.1* | | | | | *100.0* | | | | | *100.0* | | | |

Percentages of Material from Coastal Sites Categorised by Functional Context, Source, and Time Period

Table 3.13

| Context | Late Formative | | | | Early Classic | | | | | Late Classic | | | | | Terminal Classic | | | |
|---|---|---|---|---|---|---|---|---|---|---|---|---|---|---|---|---|---|---|
| | EC | SM | IX | OT | EC | SM | IX | MX | OT | EC | SM | IX | MX | OT | EC | SM | IX | OT |
| Ritual | 326 | - | - | - | 135 | 1 | 1 | - | - | 62 | - | 1 | 1 | - | - | - | - | - |
| Waste | 55 | 24 | 30 | 13 | 18 | 7 | - | - | - | 43 | 2 | 38 | - | - | 13 | - | 28 | - |
| Domestic | 7 | 1 | 2 | - | 7 | - | 2 | - | - | 14 | 1 | 12 | - | - | - | - | - | - |
| Burial | 2 | - | - | - | 2 | - | - | - | - | - | - | - | - | - | 21 | - | 12 | - |
| Ind | 3 | 1 | - | - | - | - | - | - | - | 14 | 1 | 23 | - | - | 29 | 1 | 16 | - |
| *Total 1* | *393* | *26* | *32* | *13* | *162* | *8* | *3* | - | - | *133* | *4* | *74* | *1* | - | *63* | *1* | *56* | - |
| *Total 2* | *464* | | | | *173* | | | | | *212* | | | | | *120* | | | |

Quantity of Material from All River Sites Categorised by Functional Context, Source, and Time Period

Table 3.14

| Context | Late Formative | | | | Early Classic | | | | | Late Classic | | | | | Terminal Classic | | | |
|---|---|---|---|---|---|---|---|---|---|---|---|---|---|---|---|---|---|---|
| | EC | SM | IX | OT | EC | SM | IX | MX | OT | EC | SM | IX | MX | OT | EC | SM | IX | OT |
| Ritual | 70.3 | - | - | - | 78.0 | 0.6 | 0.6 | - | - | 29.2 | - | 0.5 | 0.5 | - | - | - | - | - |
| Waste | 11.8 | 5.2 | 6.5 | 2.8 | 10.4 | 4.0 | - | - | - | 20.3 | 0.9 | 17.9 | - | - | 10.8 | - | 23.3 | - |
| Domestic | 1.5 | 0.2 | 0.4 | - | 4.0 | - | 1.2 | - | - | 6.6 | 0.5 | 5.7 | - | - | - | - | - | - |
| Burial | 0.4 | - | - | - | 1.2 | - | - | - | - | - | - | - | - | - | 17.5 | - | 10.0 | - |
| Ind | 0.6 | 0.2 | - | - | - | - | - | - | - | 6.6 | 0.5 | 10.8 | - | - | 24.2 | 0.8 | 13.3 | - |
| *Total 1* | *84.6* | *5.6* | *6.9* | *2.8* | *93.6* | *4.6* | *1.8* | - | - | *62.7* | *1.9* | *34.9* | *0.5* | - | *52.5* | *0.8* | *46.6* | - |
| *Total 2* | *99.9* | | | | *100.0* | | | | | *100.0* | | | | | *99.9* | | | |

Percentages of Material from All River Sites Categorised by Functional Context, Source, and Time Period

Table 3.15

| Context | Late Formative | | | | Early Classic | | | | | Late Classic | | | | | Terminal Classic | | | |
|---|---|---|---|---|---|---|---|---|---|---|---|---|---|---|---|---|---|---|
| | EC | SM | IX | OT | EC | SM | IX | MX | OT | EC | SM | IX | MX | OT | EC | SM | IX | OT |
| Ritual | - | - | - | - | - | - | - | - | - | - | - | - | - | - | - | - | - | - |
| Waste | 17 | 1 | 30 | 13 | 10 | 6 | - | - | 1 | 24 | - | 36 | - | - | 13 | - | 28 | - |
| Domestic | - | - | - | - | - | - | - | - | - | - | 1 | 12 | - | - | - | - | - | - |
| Burial | - | - | - | - | - | - | - | - | - | - | - | - | - | - | 21 | - | 12 | - |
| Ind | - | - | - | - | - | - | - | - | - | 11 | - | 22 | - | - | 29 | 1 | 16 | - |
| *Total 1* | *17* | *1* | *30* | *13* | *10* | *6* | - | - | *1* | *35* | *1* | *70* | - | - | *63* | *1* | *56* | - |
| *Total 2* | *61* | | | | *17* | | | | | *106* | | | | | *120* | | | |

Quantity of Material from River Sites, Excluding Blue Creek, Categorised by Functional Context, Source, and Time Period

Table 3.16

| Context | Late Formative | | | | Early Classic | | | | | Late Classic | | | | | Terminal Classic | | | |
|---|---|---|---|---|---|---|---|---|---|---|---|---|---|---|---|---|---|---|
| | EC | SM | IX | OT | EC | SM | IX | MX | OT | EC | SM | IX | MX | OT | EC | SM | IX | OT |
| Ritual | - | - | - | - | - | - | - | - | - | - | - | - | - | - | - | - | - | - |
| Waste | 27.9 | 1.6 | 49.2 | 21.3 | 58.8 | 35.3 | - | - | 5.9 | 22.6 | - | 34.0 | - | - | 10.8 | - | 23.3 | - |
| Domestic | - | - | - | - | - | - | - | - | - | - | 0.9 | 11.3 | - | - | - | - | - | - |
| Burial | - | - | - | - | - | - | - | - | - | - | - | - | - | - | 17.5 | - | 10.0 | - |
| Ind | - | - | - | - | - | - | - | - | - | 10.4 | - | 20.7 | - | - | 24.2 | 0.8 | 13.3 | - |
| *Total 1* | *27.9* | *1.6* | *49.2* | *21.3* | *58.8* | *35.3* | - | - | *5.9* | *33.0* | *0.9* | *66.0* | - | - | *52.5* | *0.8* | *46.6* | - |
| *Total 2* | *100.0* | | | | *100.0* | | | | | *99.9* | | | | | *99.9* | | | |

Percentages of Material from River Sites, Excluding Blue Creek, Categorised by Functional Context, Source, and Time Period

Table 3.17

| Context | Late Formative | | | Early Classic | | | | Late Classic | | | |
|---|---|---|---|---|---|---|---|---|---|---|---|
| | EC | SM | IX | EC | SM | IX | MX | EC | SM | IX | MX |
| Ritual | 326 | - | - | 135 | 1 | 1 | - | 62 | - | 1 | 1 |
| Waste | 38 | 23 | - | 8 | 1 | - | - | 19 | 2 | 2 | - |
| Domestic | 7 | 1 | 2 | 7 | - | 2 | - | 14 | - | - | - |
| Burial | 2 | - | - | 2 | - | - | - | - | - | - | - |
| Ind | 3 | 1 | - | - | - | - | - | 3 | 1 | 1 | - |
| *Total 1* | *376* | *25* | *2* | *152* | *2* | *3* | - | *98* | *3* | *4* | *1* |
| *Total 2* | *403* | | | *157* | | | | *106* | | | |

Quantity of Material from Blue Creek Categorised by Functional Context, Source, and Time Period

Table 3.18

| Context | Late Formative | | | Early Classic | | | | Late Classic | | | |
|---|---|---|---|---|---|---|---|---|---|---|---|
| | EC | SM | IX | EC | SM | IX | MX | EC | SM | IX | MX |
| Ritual | 81.3 | - | - | 86.0 | 0.6 | 0.6 | - | 58.5 | - | 0.9 | 0.9 |
| Waste | 9.5 | 5.7 | - | 5.1 | 0.6 | - | - | 17.9 | 1.9 | 1.9 | - |
| Domestic | 1.7 | 0.2 | 0.5 | 4.4 | - | 1.3 | - | 13.2 | - | - | - |
| Burial | 0.5 | - | - | 1.3 | - | - | - | - | - | - | - |
| Ind | 0.7 | 0.2 | - | - | - | - | - | 2.8 | 0.9 | 0.9 | - |
| *Total 1* | *93.7* | *6.1* | *0.5* | *96.8* | *1.2* | *1.9* | - | *92.4* | *2.8* | *3.7* | *0.9* |
| *Total 2* | *100.3* | | | *99.9* | | | | *99.8* | | | |

Percentages of Material from Blue Creek Categorised by Functional Context, Source, and Time Period

Table 3.19

### Discussion of Archaeological Contexts

As noted in the previous section, there are variations in the distibution of different types of obsidian. Material from El Chayal appears in more contexts than that from other sources, and is the predominant and in some cases exclusive material found in ritual deposits. While the previous section allows us to identify the various functional contexts where the material may be found, it now behooves us to examine this pattern in closer detail. This section will examine the various archaeological contexts from which the material was recovered. By further dissecting the pattern discovered in the above section, it is hoped that we may achieve a deeper understanding of the distribution mechanisms and consumption patterns that affected obsidian utilisation.
Contexts identified for this study include cache (C), construction fill, (CF), collapse (COL), floor fill (FF), floor surface (FS), humus (H), humus/mixed (H/MIX), midden (M), mortuary good (MG), other (OTH), plough-zone/surface (PZ/S), special deposit (SD), tomb shaft (TS), unknown (UNK), and workshop (WS). Explanations and definitions of these contexts may be found in Section 1 of Chapter 2.

### Middle Formative Deposits

Only the Central Peten (Tables 3.26 and 3.27) and the Inland region (Tables 3.24 and 3.25) reported Middle Formative deposits. Material from both areas is limited to construction fill contexts. It is probable that this material was originally deposited in different contexts. Rice notes that some of the 'construction fill' from the Central Peten Lakes Project included or was mixed with midden deposits (Rice 1984; Rice *et al.* 1985). The majority of this material, 71.6 per cent, was San Martin Jilotepeque obsidian, with El Chayal obsidian accounting for 17.9 per cent of the collection. Two other types of obsidan, Ixtepeque and 'Other', were also present although in lesser quantities (Table 3.21). Of note here is that none of the material reported from these regions included objects from either caches or burials (Tables 3.20 and 3.21).

### Late Formative Deposits

Material from this period is distributed throughout a wider variety of deposits than in the previous period. Objects manufactured from El Chayal obsidian are found in more contexts than the other three types, suggesting that it was sought for, and distributed to, a wider variety of uses than the other obsidian sources. That El Chayal is the exclusive source utilised in cache and mortuary deposits suggests that a preference for this source existed. Material from other sources clusters in construction fill and midden contexts, suggesting that the final disposition of these materials was not of concern to the community. Objects manufactured from Ixtepeque and San Martin Jilotepeque also appear in floor contexts, either as fill or on the floor surfaces, implying that these sources were considered suitable for domestic functions. El Chayal material also occurrs in these contexts, but in lower quantities than in cache contexts.
Artefacts recovered from the Coastal and Inland regions were limited to construction fill contexts (Tables 3.22 to 3.25). While the sites in the Central Peten and River areas had a greater diversity of contexts, construction fill still dominated the consumption pattern. In the Central Peten the largest concentration of obsidian was recovered from this context (Tables 3.26 and 3.27), while among River sites it was the second most common location, caches being the first (Tables 3.28 and 3.29)

### Early Classic Deposits

The pattern for obsidian consumption in the Early Classic closely mirrors that of the preceeding period. While El Chayal is no longer the exclusive source utilised in caches, it remains predominant, contributing 97.4 per cent of the objects used in these deposits. Small amounts of Mexican (1.5 per cent), Ixtepeque, and San Martin Jilotepeque (0.5 per cent each) material are also found in caches. Material from the latter two sources amounts to single pieces from each source. The Coastal area was the only location for which caches were not reported in this period. In the River and Inland areas, El Chayal material dominates these deposits. Inland caches contain both El Chayal and small amounts of Mexican material (5.6 per cent), while River sites counted the single pieces of San Martin and Ixtpeque material mentioned above as part of their cache deposits. Central Peten cache material was limited to three pieces of El Chayal, with no material from other sites being reported.

As with the previous period, the majority of San Martin Jilotepeque and Ixtepeque material is concentrated in construction fill and midden contexts, with limited amounts of Ixtepeque material in domestic contexts. Regionally, San Martin Jilotepeque material is only reported in the Central Peten and River areas during this period. In both of these areas, San Martin Jilotepeque material is found predominantly in construction fill contexts, with a small amount being recovered in middens at Peten sites. Ixtepeque material was recovered in River, Inland, and Coastal sites.

The introduction of Mexican obsidian during this period creates an interesting ripple in the pattern of obsidian consuption patterns. As this material comes from greater distances, it has been hypothesised that this material would be more highly coveted, and therefore utilised in status deposits. While Mexican material does appear in cache and mortuary contexts (n=3 each), the majority of the material appears in construction fill or structural collapse (n=11). Mexican obsidian is present in both the Central Peten and Inland areas. The material in the Central Peten was recovered primarily from construction contexts, although moderate amounts of the material were distributed in midden, humus, domestic, and burial contexts. No Mexican obsidian was reported from Peten caches. In the Inland area, Mexican material was recovered only from, cache and construction fill context.

Caches form the largest type of deposit in the Inland area with the only other deposit type being construction fill (Tables 3.24 and 3.25). A similar pattern is present in the River area where caches are also the most common type of deposit and construction fill being the second most common context (Tables 3.28 and 3.29). In the Central Peten, while construction fill deposits yeilded the largest concentrations of obsidian, with 57.4 per cent of the collection, caches contained a meager 5.6 per cent of the collection (Tables 3.26 and 3.27). However, this is still larger than the Coastal area where no caches were reported and the collection was distributed roughly evenly between construction fill, midden, and mortuary deposits (Tables 3.23 and 3.23).

### LATE CLASSIC DEPOSITS

The most distinct shift in obsidian consumption in the Late Classic is away from caches. A new category, Special Deposit, was introduced for material recovered from this period. These deposits, while ritualistic in nature, defy the classic definition of a cache in that the material is not collected from a single, concentrated location, but spread over or around an area (see Chapter 2 Section 1). Caches consumed only 3.7 per cent of the Late Classic collection, while material in Special Deposits absorbed 11.3 per cent of the material. If these two deposits are considered ideologically equivalent, then ceremonial activities still account for only 14.6 per cent of obsidian consumption. This is significantly lower then the 58 per cent documented for the Late Formative and the 60.7 per cent known from the Early Classic (Tables 3.20 and 3.21).

In the Early Classic caches had previously been reported in three of the four geographic areas, River, Central Peten and Inland. During the Late Classic, no caches were reported in the Inland area, when they had previously accounted for 71.0 per cent of the obsidian consumed. In the River area the decrease in obsidian consumption in caches was almost as dramatic, falling from 78.8 per cent in the Early Classic to 7.1 per cent in the Late Classic. Declines in Central Peten cache consumption were still documented, albeit less excitingly, from 5.6 per cent in the Early Classic to 0.9 per cent in the Late Classic.

A second type of deposit, Tomb Shafts, is introduced in this period, appearing in the Central Peten and Inland areas (Tables 3.24 to 3.27). As discussed in the preceeding chapter, these deposits are problematic as the nature of the offerings (public, private, ceremonial or personal grave goods) is unclear. However, it is apparent that these deposit are ritual in function. Tomb Shafts account for 21.2 per cent of the obsidian consumed in this period. When this amount is tallied with caches and special deposits, the total ritual activities still amount to only 35.8 per cent of the collection. As noted previously in the discussion of functional contexts, this is the largest type of deposit for this period. However, it is still substantially smaller than in previous periods.

Of interest for this study is that, regardless of the shift in archaeological contexts, the primary if not exclusive source utilised is El Chayal. All cache and tomb shaft material identified belong to this source, and 95.9 per cent of material identified in special deposits was also El Chayal in origin.

San Martin Jilotepeque material remains restricted primarily to middens and fill deposits. The largest concentration of this material is from construction fill contexts (3.0 per cent of the Late Classic collection, 70 per cent of the San Martin Jilotepeque material). A small percentage of material was recovered from domestic contexts.

Ixtepeque material shows little variation from the previous periods. The majority of material is still found in construction fill contexts, with smaller amounts distributed in midden and domestic contexts. Where the distribution of this material changes is with the addition of small amounts of this material being recovered from special deposits and workshops.

Mexican material is still present in this period, albeit in lower numbers (n-5) than in the previous periods (n=23). Distribution of this material is reduced to three contexts: construction fill, which accounts for the largest amount of Mexican material (0.7 per cent of the Late Classic collection), and humus and special deposits, which each possessed only 0.2 per cent of the Late Classic collection.

### TERMINAL CLASSIC DEPOSITS

Deposits in the Terminal Classic period from these two areas decrease in variety. Material was recovered from four identifiable deposits: construction fill, middens, mortuary goods, and workshops. As with the previous period, the largest type of deposit was construction fill, which accounted for 26.1 per cent of the Terminal Classic collection Tables 3.20 and 3.21). The next largest identifiable deposit was mortuary goods, which accounted for 23.4 per cent of the collection. Only El Chayal and Ixtepeque were utilised for this purpose. Both of these sources were utilised in the same series of contexts, construction fill midden, mortuary goods, and workshops. San Martin Jilotepeque material was recovered from construction fill. As has been identified, Ixtepeque material becomes more prevalent during this period. That it appears concurrent to El Chayal sources suggests that it may have been replacing El Chayal as the material of choice for important deposits.

While material from this period was recovered in the Coastal, Central Peten and River areas, the Coastal collection was too small to allow for any pattern to be accurately noted (Tables 3.22 and 3.23). As in the previous period the majority of Central Peten obsidian was recovered from construction fill contexts, the only other context noted in this area in this period was midden (Tables 3.26 and 3.27). Deposits in the River area were slightly more varied (Tables 3.28 and 3.29). Mortuary goods were the largest proportion of obsidian in a securely identifiable context (27.3 per cent), followed by workshops (17.3 per cent), and construction fill (16.5 per cent) (Tables 3.28 and 3.29).

### Summary of Archaeological Contexts

Examination of obsidian consumption by archaeological context reveals that El Chayal material is predominant in cache contexts in the Late Formative and Early Classic periods. During the Late Classic, ritual activities appear to shift away from caching in favour of special deposits and tomb shafts. Terminal Classic patterns eschew these activities completely, and the majority of material in identifiable contexts is recovered from construction fill, with mortuary goods representing the only remotely ritualistic activity. Material from the Middle Formative is limited to construction fill contexts. However, this may be due to limited sampling or recycling of deposits by later periods, and may not represent an accurate depiction of Middle Formative obsidian distribution.

El Chayal material is the most widely distributed material from the Late Formative through the Late Classic. The majority of this material is concentrated in caches or in the Late Classic special deposits and tomb shafts.

San Martin Jilotepeque material appears predominantly in construction fill and midden contexts. A very small amount (n=1) appears in cache context in the Early Classic. However, this piece appears to be the exception rather than the norm. Material from this source appears not to have been considered for inclusion in important or ritual deposits.

Objects manufactured from Ixtepeque material were recovered from deposits similar to San Martin Jilotepeque items. Differences in the deposits occur in the Late Classic, when Ixtepeque material was found in special deposits. Divergence in the consumption pattern of Ixtepeque and San Martin Jilotepeque material is continued in the Terminal Classic, when the contextual pattern of Ixtepeque material more closely mimics El Chayal.

Mexican material is introduced in the Early Classic period, where it appears predominantly in construction fill contexts. Items manufactured from this material were also found in mortuary contexts and caches, albeit in more limited amounts. Small amounts were also recovered from midden contexts. The Late Classic pattern is similar, with material appearing in special deposits instead of caches.

Hypotheses regarding obsidian consumption may note that while El Chayal material appears to be the material of choice for important or ritual activities, it still 'trickles down' to other, less prestigeous deposits. The opposite is true for San Martin Jilotepeque material. That this material appears restricted to less or non-prestigeous contexts, and almost never filters up. That suggests that El Chayal was considered the more important or preferred material, and that it was desired for a variety of functions. San Martin Jilotepeque material, on the other hand was not so widely coveted, and used when El Chayal or other more preferred material could not be had. Material from the Ixtepeque source was not as valued as El Chayal and initially appears to have been considered on par with San Martin Jilotepeque material. Items of Mexican material were more broadly viewed and appear in contexts similar to El Chayal, but in smaller degrees.

Although the nuances of these patterns can be quite subtle, the broad impression is that the sources are very clearly tied to preference and archaeological contexts. The next question that needs to be addressed is whether these preferenial patterns are also linked to status of consumers. The next section will address this question.

### Discussions of Social Contexts

It has already been established in the preceding text that an association exists between source utilisation an functional and archaeological contexts. This analytical section is designed to determine if a similar correlation existed between source utilisation and social status. Social contexts were determined based on location of the deposits and associated architecture. Nine categories were created, Elite 1, Elite 2, Non-Elite 1, Non-Elite 2, Non-Elite 3, Ritual/Ceremonial 1, Ritual Ceremonial 2, Workshop and Indeterminate. A full description of these contexts may be found in Chapter 2 Section 1.

### Middle Formative Deposits

As material from this period was recovered from construction fill or mixed construction fill/midden contexts it was deemed infeasible to determine the social origin of the material. Consequently contexts for this material was identified as 'indeterminate' (Tables 3.34 and 3.35). Material from this period was recovered only in the Inland and Central Peten areas (Tabels 3.38 and 3.39, and Tables 3.40 and 3.41 repectively).

### LATE FORMATIVE DEPOSITS

Obsidian from this period was recovered from four different social contexts; Non-Elite 1, Non-Elite 2, Ritual/Ceremonial 1, and Indeterminate (Tables 3.34 and 3.35). The largest concentration of material was recovered from the Ritual/ Ceremonial context. Concentrations of obsidian in these deposits accounted for 57.8 per cent of the Late Formative collection. Non-Elite 1 deposits amount to a total of 12.1 per cent of the collection. Although identified here as Non-Elites it is probable that individuals in this moiety represented the highest status group at this point in time. Non-Elite 2 deposits utilised substantially lower quantities of material, only 1.6 per cent of the collection.

Source utilisation for this period demonstrates that both Non-Elite contexts were utilising El Chayal and San Martin Jilotepeque material while only El Chayal material appears in the Ritual/Ceremonial 1 context. The presence of only El Chayal material in ritual and cache contexts was already established in the preceding sections on Functional and Archaeological contexts. What is of interest is that there appears to be relatively little differentiation between consumption of El Chayal and San Martin Jilotepeque material among the Non-Elite 1 consumers, with the obsidian sources being split 52.9 per cent and 47.1 per cent respectively. This pattern changes dramatically in the following period.

### EARLY CLASSIC DEPOSITS

During this period distribution of obsidian changes radically from that of the Late Formative. Most noticeably is the drop in Ritual Ceremonial deposits to 18.4 per cent from 57.8 per cent. Consumption of obsidian by the highest status moiety increases dramatically and the Elite 1 deposits account for 43.9 per cent of the material. This context is the largest consumer of obsidian for this period, out-weighing even the 'indeterminate' deposits at 31.9 per cent. As noted previously the Non-Elite 1 deposits of the Late Formative may be considered roughly analogous to the Elite deposits of this period as they both represent the highest social group. Comparisons of these 'highest status' deposits reveals a substantial jump in the per cent of material being utilised in these deposits.

Following on the above concept the Non-Elite 2 context of the preceding period may be considered analogous to the Non-Elite 1 and 2 contexts of the Early Classic period. Non-Elite deposits in the Early Classic period accounted for 5.6 per cent of obsidian for this period. The jump in consumption by this group of people is almost identical to that in the Elite contexts, 350 per cent for the Non-Elites and 360 per cent for the Elites.

El Chayal material is found in all contexts during this period, while San Martin Jilotepeque is found almost exclusively in Non-Elite 1 and Indeterminate contexts (Tables 3.34 and 3.35). A single piece of San Martin Jilotepeque obsidian was found in an Elite 1 context. No pieces were found in either Elite 2 or Ritual/Ceremonial contexts.

### LATE CLASSIC DEPOSITS

Declines in Ritual/Ceremonial 1 contexts continue in the Late Classic period where deposits belong to this category total only 0.2 per cent of the collection. Ritual/Ceremonial 2 contexts begin during this period. These contexts, as outline in Chapter 2 Section 1, are still 'public' in location, but problematic in terms of determining the social origin or initial ownership of the material. These deposits account for 23.1 per cent of the obsidian from this period. Elite consumption declines slightly to 36.0 per cent. This amount is divided roughly equally between the Elite 1 and the Elite 2 contexts with 18.1 and 17.9 per cent of the collection respectively. Non-Elite deposits increase substantially in this period almost tripling from 5.6 per cent to 14.6 per cent. While material from Workshops appears for the first time.

As with the preceding periods, San Martin Jilotepeque material is not found in Ritual/Ceremonial contexts. Curiously, this material is also not found in Workshop deposits. Objects from both these contexts, Ritual/Ceremonial and Workshop, are limited to El Chayal and Ixtepeque sources. San Martin Jilotepeque material is found in both Elite and Non-Elite contexts although the material present in Elite contexts is considerably smaller than in Non-Elite contexts. In Elite contexts objects manufactured from San Martin Jilotepeque obsidian account for only 1.5 per cent of the Elite collection, while they account for 5.2 per cent of the Non-Elite material.

### TERMINAL CLASSIC DEPOSITS

Consumption of material by Elites during this period remains fairly consistent with levels established during the Late Classic Period. In the preceding period this amount equalled 36.0 per cent while in this period the amount equals 34.0 per cent. Non-Elite deposits drop to 0.7 per cent of the collection. However this may be due to a sampling bias or the limited reporting using this

| Context | Middle Formative | | | | Late Formative | | | | Early Classic | | | | | Late Classic | | | | | Terminal Classic | | | |
|---|---|---|---|---|---|---|---|---|---|---|---|---|---|---|---|---|---|---|---|---|---|---|
| | EC | SM | IX | OT | EC | SM | IX | OT | EC | SM | IX | MX | OT | EC | SM | IX | MX | OT | EC | SM | IX | OT |
| C | - | - | - | - | 326 | - | - | - | 189 | 1 | 1 | 3 | - | 16 | - | - | - | - | - | - | - | - |
| CF | 12 | 48 | 5 | 2 | 56 | 28 | 31 | 13 | 56 | 7 | 9 | 8 | 2 | 55 | 13 | 24 | 3 | 3 | 17 | 1 | 16 | 3 |
| COL | - | - | - | - | 1 | - | - | - | 1 | - | - | 3 | - | 2 | - | - | - | - | - | - | - | - |
| FF | - | - | - | - | 2 | 4 | 1 | - | 7 | - | 1 | - | - | 4 | 1 | 12 | - | - | - | - | - | |
| FS | - | - | - | - | 12 | - | 1 | - | - | - | 1 | 1 | - | 13 | - | - | - | - | - | - | - | - |
| H | - | - | - | - | 4 | 1 | - | - | - | - | - | 4 | - | 4 | - | - | 1 | - | - | - | - | - |
| H/F | - | - | - | - | - | - | - | - | - | - | - | - | - | 8 | 2 | 1 | - | - | - | - | - | - |
| M | - | - | - | - | 37 | 33 | 1 | - | 8 | 2 | 1 | 1 | - | 58 | 3 | 12 | - | - | 1 | - | 2 | - |
| MG | - | - | - | - | 3 | - | - | - | 8 | - | 1 | 3 | 2 | - | - | - | - | - | 21 | - | 12 | - |
| OTH | - | - | - | - | 2 | - | - | - | - | - | - | - | - | 2 | 1 | 1 | - | - | - | - | - | - |
| PZ/S | - | - | - | - | 1 | 1 | - | - | - | - | - | - | - | 1 | - | - | - | - | - | - | - | - |
| SD | - | - | - | - | - | - | - | - | - | - | - | - | - | 47 | - | 1 | 1 | - | - | - | - | - |
| TS | - | - | - | - | - | - | - | - | - | - | - | - | - | 91 | - | - | - | - | - | - | - | - |
| UNK | - | - | - | - | - | - | - | - | - | - | - | - | - | 11 | - | 22 | - | - | 29 | 1 | 17 | - |
| WS | - | - | - | - | 1 | - | 3 | - | - | - | - | - | - | 9 | - | 8 | - | - | 9 | - | 12 | - |
| *Total 1* | *12* | *48* | *5* | *2* | *445* | *67* | *37* | *13* | *269* | *10* | *14* | *23* | *4* | *321* | *20* | *81* | *5* | *3* | *77* | *2* | *59* | *3* |
| *Total 2* | *67* | | | | *562* | | | | *320* | | | | | *430* | | | | | *141* | | | |

Quantity of Material from All Sites Categorised by Archaeological Context, Source, and Time Period

Table 3.20

| Context | Middle Formative | | | | Late Formative | | | | Early Classic | | | | | Late Classic | | | | | Terminal Classic | | | |
|---|---|---|---|---|---|---|---|---|---|---|---|---|---|---|---|---|---|---|---|---|---|---|
| | EC | SM | IX | OT | EC | SM | IX | OT | EC | SM | IX | MX | OT | EC | SM | IX | MX | OT | EC | SM | IX | OT |
| C | | | | | 58.0 | - | - | - | 59.2 | 0.3 | 0.3 | 0.9 | - | 3.7 | - | - | - | - | - | - | - | - |
| CF | 17.9 | 71.6 | 7.5 | 3.0 | 10.0 | 5.0 | 5.5 | 2.3 | 17.5 | 2.2 | 2.8 | 2.5 | 0.6 | 12.8 | 3.0 | 5.6 | 0.7 | 0.7 | 12.0 | 0.7 | 11.3 | 2.1 |
| COL | - | - | - | - | 0.2 | - | - | - | 0.3 | - | - | 0.9 | - | 0.5 | - | - | - | - | - | - | - | - |
| FF | - | - | - | - | 0.4 | 0.7 | 0.2 | - | 2.2 | - | 0.3 | - | 0.3 | 0.9 | 0.2 | 2.8 | - | - | - | - | - | - |
| FS | - | - | - | | 2.1 | - | 0.2 | - | - | - | 0.3 | 0.3 | - | 0.3 | - | - | - | - | - | - | - | - |
| H | - | - | - | - | 0.7 | 0.2 | - | - | - | - | - | 1.2 | - | 0.9 | - | - | 0.2 | - | - | - | - | - |
| H/F | - | - | - | - | - | - | - | - | - | - | - | - | - | 1.9 | 0.5 | 0.2 | - | - | - | - | - | - |
| M | - | - | - | - | 6.6 | 5.9 | 0.2 | - | 2.5 | 0.6 | 0.3 | 0.3 | - | 13.5 | 0.7 | 2.8 | - | - | 0.7 | - | 1.4 | - |
| MG | - | - | - | - | 0.5 | - | - | - | 2.5 | - | 0.3 | 0.9 | 0.6 | - | - | - | - | - | 14.9 | - | 8.5 | - |
| OTH | - | - | - | - | 0.4 | - | - | - | - | - | - | - | - | 0.5 | 0.2 | 0.2 | - | - | - | - | - | - |
| PZ/S | - | - | - | - | 0.2 | 0.2 | - | - | - | - | - | - | - | 0.2 | - | - | - | - | - | - | - | - |
| SD | - | - | - | - | - | - | - | - | - | - | - | - | - | 10.9 | - | 0.2 | 0.2 | - | - | - | - | - |
| TS | - | - | - | - | - | - | - | - | - | - | - | - | - | 21.2 | - | - | - | - | - | - | - | - |
| UNK | - | - | - | - | - | - | - | - | - | - | - | - | - | 2.6 | - | 5.1 | - | - | 20.6 | 0.7 | 12.0 | - |
| WS | - | - | - | - | 0.2 | - | 0.5 | - | - | - | - | - | - | 2.1 | - | 1.9 | - | - | 6.4 | - | 8.5 | - |
| *Total 1* | *17.9* | *71.6* | *7.5* | *3.0* | *79.3* | *12.0* | *6.1* | *2.3* | *84.2* | *3.1* | *4.0* | *7.0* | *1.5* | *74.7* | *4.6* | *18.8* | *1.1* | *0.7* | *54.6* | *1.4* | *41.7* | *2.1* |
| *Total 2* | *100.0* | | | | *99.7* | | | | *100.1* | | | | | *99.9* | | | | | *99.8* | | | |

Percentages of Material from All Sites Categorised by Archaeological Context, Source, and Time Period

Table 3.21

| Context | Late Formative | | | Early Classic | | | Late Classic | | | Terminal Classic | | |
|---|---|---|---|---|---|---|---|---|---|---|---|---|
| | ELC | SMJ | IXT | ELC | SMJ | IXT | ELC | SMJ | IXT | ELC | SMJ | IXT |
| CF | 16 | - | - | 3 | - | 1 | - | - | - | 1 | - | - |
| M | - | - | - | 6 | - | - | 17 | - | 3 | - | - | - |
| MG | - | - | - | 4 | - | 1 | - | - | - | - | - | - |
| *Total 1* | *16* | - | - | *13* | - | *2* | *17* | - | *3* | *1* | - | - |
| *Total 2* | *16* | | | *15* | | | *20* | | | *1* | | |

Quantity of Coastal Material Categorised by Archaeological Context, Source, and Time Period

Table 3.22

| Context | Late Formative | | | Early Classic | | | Late Classic | | | Terminal Classic | | |
|---|---|---|---|---|---|---|---|---|---|---|---|---|
| | ELC | SMJ | IXT | ELC | SMJ | IXT | ELC | SMJ | IXT | ELC | SMJ | IXT |
| CF | 100 | - | - | 20.0 | - | 6.7 | - | - | - | 100.0 | - | - |
| M | - | - | - | 40.0 | - | - | 85.0 | - | 15.0 | - | - | - |
| MG | - | - | - | 26.7 | - | 6.7 | - | - | - | - | - | - |
| *Total 1* | *100* | - | - | *80.7* | - | *13.4* | *85.0* | - | *15.0* | *100.0* | - | - |
| *Total 2* | *100.0* | | | *100.1* | | | *100.0* | | | *100.0* | | |

Percentages of Coastal Material Categorised by Archaeological Context, Source, and Time Period

Table 3.23

| Context | Middle Formative | | | Late Formative | | | Early Classic | | | | Late Classic | | |
|---|---|---|---|---|---|---|---|---|---|---|---|---|---|
| | ELC | SMJ | IXT | ELC | SMJ | IXT | ELC | SMJ | IXT | MX | ELC | SMJ | IXT |
| C | - | - | - | - | - | - | 51 | - | - | 3 | - | - | - |
| CF | 1 | 1 | 2 | 1 | - | 1 | 18 | - | 2 | 2 | 1 | - | - |
| TS | - | - | - | - | - | - | - | - | - | - | 86 | - | - |
| *Total 1* | *1* | *1* | *2* | *1* | - | *1* | *69* | - | *2* | *5* | *87* | - | - |
| *Total 2* | *4* | | | *2* | | | *76* | | | | *87* | | |

Quantity of Material from Inland Sites Categorised by Archaeological Context, Source, and Time

Table 3.24

| Context | Middle Formative | | | Late Formative | | | Early Classic | | | | Late Classic | | |
|---|---|---|---|---|---|---|---|---|---|---|---|---|---|
| | ELC | SMJ | IXT | ELC | SMJ | IXT | ELC | SMJ | IXT | MX | ELC | SMJ | IXT |
| C | - | - | - | - | - | - | 67.1 | - | - | 3.9 | - | - | - |
| CF | 25 | 25 | 50 | 50 | - | 50 | 23.7 | - | 2.6 | 2.6 | 1.1 | - | - |
| TS | - | - | - | - | - | - | - | - | - | - | 98.8 | - | - |
| *Total 1* | *25* | *25* | *50* | *50* | - | *50* | *90.8* | - | *2.6* | *6.5* | *99.9* | - | - |
| *Total 2* | *100* | | | *100* | | | *99.9* | | | | *99.9* | | |

Percentages of Material from Inland Sites Categorised by Archaeological Context, Source and Time

Table 3.25

| Context | Middle Formative | | | | Late Formative | | | Early Classic | | | | EC/ LC | Late Classic | | | | | Terminal Classic | | | |
|---|---|---|---|---|---|---|---|---|---|---|---|---|---|---|---|---|---|---|---|---|---|
| | EC | SM | IX | OT | EC | SM | IX | EC | SM | MX | OT | MX | EC | SM | IX | MX | OT | EC | SM | IX | OT |
| C | - | - | - | - | - | - | - | 3 | - | - | - | - | 1 | - | - | - | - | - | - | - | - |
| CF | 11 | 47 | 3 | 2 | 15 | 27 | 4 | 18 | 6 | 6 | 1 | 3 | 40 | 13 | 4 | - | 3 | 12 | 1 | - | 3 |
| COL | - | - | - | - | 1 | - | - | - | - | 3 | - | - | 1 | - | - | - | - | - | - | - | - |
| FF | - | - | - | - | 1 | 3 | - | - | - | - | - | - | 3 | - | - | - | - | - | - | - | - |
| FS | - | - | - | - | 6 | - | - | - | - | 1 | - | - | - | - | - | - | - | - | - | - | - |
| H | - | - | - | - | 4 | 1 | - | - | - | 4 | - | - | - | - | - | 1 | - | - | - | - | - |
| M | - | - | - | - | 6 | 10 | - | 2 | 2 | 1 | - | - | 34 | 3 | - | - | - | 1 | - | 2 | - |
| MG | - | - | - | - | 1 | - | - | 2 | - | 3 | 2 | - | - | - | - | - | - | - | - | - | - |
| TS | - | - | - | - | - | - | - | - | - | - | - | - | 5 | - | - | - | - | - | - | - | - |
| WS | - | - | - | - | 1 | - | - | - | - | - | - | - | - | - | - | - | - | - | - | - | - |
| *Total 1* | *11* | *47* | *3* | *2* | *35* | *41* | *4* | *25* | *8* | *18* | *3* | *3* | *84* | *16* | *4* | *1* | *3* | *13* | *1* | *2* | *3* |
| *Total 2* | *63* | | | | *80* | | | *54* | | | | *3* | *108* | | | | | *19* | | | |

Quantity of Material from Peten Sites Categorised by Archaeological Context, Source, and Time Period

Table 3.26

| Context | Middle Formative | | | | Late Formative | | | Early Classic | | | | EC/LC | Late Classic | | | | | Terminal Classic | | | |
|---|---|---|---|---|---|---|---|---|---|---|---|---|---|---|---|---|---|---|---|---|---|
| | EC | SM | IX | OT | EC | SM | IX | EC | SM | MX | OT | MX | EC | SM | IX | MX | OT | EC | SM | IX | OT |
| C | - | - | - | - | - | - | - | 5.6 | - | - | - | - | 0.9 | - | - | - | - | - | - | - | - |
| CF | 17.5 | 74.6 | 4.8 | 3.2 | 18.8 | 33.8 | 5.0 | 33.3 | 11.1 | 11.1 | 1.9 | 100 | 37.0 | 12.0 | 3.7 | - | 2.8 | 63.1 | 5.3 | - | 15.8 |
| COL | - | - | - | - | 1.3 | - | - | - | - | 5.6 | - | - | 0.9 | - | - | - | - | - | - | - | - |
| FF | - | - | - | - | 1.3 | 3.8 | - | - | - | - | - | - | 2.8 | - | - | - | - | - | - | - | - |
| FS | - | - | - | - | 7.5 | - | - | - | - | 1.9 | - | - | - | - | - | - | - | - | - | - | - |
| H | - | - | - | - | 5.0 | 1.3 | - | - | - | 7.4 | - | - | - | - | - | 0.9 | - | - | - | - | - |
| M | - | - | - | - | 7.5 | 12.5 | - | 3.7 | 3.7 | 1.9 | - | - | 31.5 | 2.8 | - | - | - | 5.3 | - | 10.5 | - |
| MG | - | - | - | - | 1.3 | - | - | 3.7 | - | 5.6 | 3.7 | - | - | - | - | - | - | - | - | - | - |
| TS | - | - | - | - | - | - | - | - | - | - | - | - | 4.6 | - | - | - | - | - | - | - | - |
| WS | - | - | - | - | 1.3 | - | - | - | - | - | - | - | - | - | - | - | - | - | - | - | - |
| *Total 1* | *17.5* | *74.6* | *4.8* | *3.2* | *44.0* | *51.4* | *5.0* | *46.3* | *14.8* | *33.5* | *5.6* | *100* | *77.7* | *14.8* | *3.7* | *0.9* | *2.8* | *68.4* | *5.3* | *10.5* | *15.8* |
| *Total 2* | *100.1* | | | | *100.4* | | | *100.2* | | | | *100* | *99.9* | | | | | | | | |

Percentages of Material from Peten Sites Categorised by Archaeological Context, Source, and Time Period

Table 3.27

| Context | Late Formative | | | | Early Classic | | | | Late Classic | | | | Terminal Classic | | |
|---|---|---|---|---|---|---|---|---|---|---|---|---|---|---|---|
| | ELC | SMJ | IXT | OTH | ELC | SMJ | IXT | OTH | ELC | SMJ | IXT | MX | ELC | SMJ | IXT |
| C | 326 | - | - | - | 135 | 1 | 1 | - | 15 | - | - | - | - | - | - |
| CF | 24 | 1 | 26 | 13 | 17 | 7 | - | 1 | 14 | - | 20 | - | 4 | - | 16 |
| COL | - | - | - | - | 1 | - | - | - | 1 | - | - | - | - | - | - |
| FF | 1 | 1 | 1 | - | 7 | - | 1 | - | 1 | 1 | 12 | - | - | - | - |
| FS | 6 | - | 1 | - | - | - | 1 | - | 13 | - | - | - | - | - | - |
| H | - | - | - | - | - | - | - | - | 4 | - | - | - | - | - | - |
| H/MIX | - | - | - | - | - | - | - | - | 8 | 2 | 1 | - | - | - | - |
| M | 31 | 23 | 1 | - | - | - | - | - | 7 | - | 9 | - | - | - | - |
| MG | 2 | - | - | - | 2 | - | - | - | - | - | - | - | 21 | - | 12 |
| OTH | 2 | - | - | - | - | - | - | - | 13 | 1 | 23 | - | 29 | 1 | 17 |
| PZ/S | 1 | 1 | - | - | - | - | - | - | 1 | - | - | - | - | - | - |
| SD | - | - | - | - | - | - | - | - | 47 | - | 1 | 1 | - | - | - |
| WS | - | - | 3 | - | - | - | - | - | 9 | - | 8 | - | 9 | - | 12 |
| *Total 1* | *393* | *26* | *32* | *13* | *162* | *8* | *3* | *1* | *133* | *4* | *74* | *1* | *63* | *1* | *57* |
| *Total 2* | *464* | | | | *174* | | | | *212* | | | | *121* | | |

Quantities of Material from All River Sites Categorised by Archaeological Context, Source, and Time Period

Table 3.28

| Context | Late Formative | | | | Early Classic | | | | Late Classic | | | | Terminal Classic | | |
|---|---|---|---|---|---|---|---|---|---|---|---|---|---|---|---|
| | ELC | SMJ | IXT | OTH | ELC | SMJ | IXT | OTH | ELC | SMJ | IXT | MX | ELC | SMJ | IXT |
| C | 70.3 | - | - | - | 77.6 | 0.6 | 0.6 | - | 7.1 | - | - | - | - | - | - |
| CF | 5.2 | 0.2 | 5.6 | 2.8 | 9.8 | 4.0 | - | 0.6 | 6.6 | - | 9.4 | - | 3.3 | - | 13.2 |
| COL | - | - | - | - | 0.6 | - | - | - | 0.5 | - | - | - | - | - | - |
| FF | 0.2 | 0.2 | 0.2 | - | 4.0 | - | 0.6 | - | 0.5 | 0.5 | 5.7 | - | - | - | - |
| FS | 1.3 | - | 0.2 | - | - | - | 0.6 | - | 6.1 | - | - | - | - | - | - |
| H | - | - | - | - | - | - | - | - | 1.9 | - | - | - | - | - | - |
| H/MIX | - | - | - | - | - | - | - | - | 3.8 | 1.0 | 0.5 | - | - | - | - |
| M | 6.7 | 5.0 | 0.2 | - | - | - | - | - | 3.3 | - | 4.2 | - | - | - | - |
| MG | 0.4 | - | - | - | 0.9 | - | - | - | - | - | - | - | 17.4 | - | 9.9 |
| OTH | 0.4 | - | - | - | - | - | - | - | 6.1 | 0.5 | 10.8 | - | 23.9 | 0.8 | 14.0 |
| PZ/S | 0.2 | 0.2 | - | - | - | - | - | - | 0.5 | - | - | - | - | - | - |
| SD | - | - | - | - | - | - | - | - | 22.2 | - | 0.5 | 0.5 | - | - | - |
| WS | - | - | 0.6 | - | - | - | - | - | 4.2 | - | 3.8 | - | 7.4 | - | 9.9 |
| *Total 1* | *84.7* | *5.6* | *6.8* | *2.8* | *92.9* | *4.6* | *1.8* | *0.6* | *62.8* | *2.0* | *34.9* | *0.5* | *52.0* | *0.8* | *47.0* |
| *Total 2* | *99.9* | | | | *99.9* | | | | *100.2* | | | | *99.8* | | |

Percentages of Material from All River Sites Categorised by Archaeological Context, Source, and Time Period

Table 3.29

| Context | Late Formative | | | | Early Classic | | | Late Classic | | | Terminal Classic | | |
|---|---|---|---|---|---|---|---|---|---|---|---|---|---|
| | ELC | SMJ | IXT | OTH | ELC | IXT | OTH | ELC | SMJ | IXT | ELC | SMJ | IXT |
| CF | 17 | - | 26 | 13 | 10 | 6 | 1 | 8 | | 19 | 4 | - | 16 |
| FF | - | - | - | - | - | - | - | | 1 | 12 | - | - | - |
| M | - | 1 | 1 | - | - | - | - | 7 | | 9 | - | - | - |
| MG | - | - | - | - | - | - | - | | | - | 21 | - | 12 |
| UNK | - | - | - | - | - | - | - | 11 | | 22 | 29 | 1 | 17 |
| WS | - | - | 3 | - | - | - | - | 9 | | 8 | 9 | - | 12 |
| *Total 1* | *17* | *1* | *30* | *13* | *10* | *6* | *1* | *35* | *1* | *70* | *63* | *1* | *57* |
| *Total 2* | *61* | | | | *17* | | | *106* | | | *42* | | |

Quantities of Material from River Sites, Excluding Blue Creek, Categorised by Archaeological Context, Source, and Time Period

Table 3.30

| Context | Late Formative | | | | Early Classic | | | Late Classic | | | Terminal Classic | | |
|---|---|---|---|---|---|---|---|---|---|---|---|---|---|
| | ELC | SMJ | IXT | OTH | ELC | IXT | OTH | ELC | SMJ | IXT | ELC | SMJ | IXT |
| CF | 27.9 | - | 42.6 | 21.3 | 58.8 | 35.5 | 5.9 | 7.5 | - | 17.9 | 3.3 | - | 13.2 |
| FF | - | - | - | - | - | - | - | - | 0.9 | 11.3 | - | - | - |
| M | - | 1.6 | 1.6 | - | - | - | - | 6.6 | - | 8.5 | - | - | - |
| MG | - | - | - | - | - | - | - | - | - | - | 17.4 | - | 9.9 |
| UNK | - | - | - | - | - | - | - | 10.4 | - | 20.8 | 23.9 | 0.8 | 14.0 |
| WS | - | - | 4.9 | - | - | - | - | 8.5 | - | 7.5 | 7.4 | - | 9.9 |
| *Total 1* | *27.9* | *1.6* | *49.1* | *21.3* | *58.8* | *35.5* | *5.9* | *33.0* | *0.9* | *66.1* | *52.0* | *0.8* | *47.0* |
| *Total 2* | *99.9* | | | | *100.2* | | | *100.0* | | | *99.8* | | |

Percentages of Material from River Sites, Excluding Blue Creek, Categorised by Archaeological Context, Source, and Time Period

Table 3.31

| Context | Late Formative | | | | Early Classic | | | | Late Classic | | | | |
|---|---|---|---|---|---|---|---|---|---|---|---|---|---|
| | ELC | SMJ | IXT | NA | ELC | SMJ | IXT | NA | ELC | SMJ | IXT | NA | MX |
| C | 326 | - | - | 146 | 135 | 1 | 1 | 106 | 15 | - | - | 6 | - |
| CF | 7 | 1 | - | 11 | 7 | 1 | - | 2 | 6 | - | 1 | - | - |
| COL | - | - | - | - | 1 | - | - | - | 1 | - | - | - | - |
| FF | 1 | 1 | 1 | 1 | 7 | - | 1 | 2 | 1 | - | - | - | - |
| FS | 6 | - | 1 | 1 | - | - | 1 | - | 13 | - | - | 1 | - |
| H | - | - | - | - | - | - | - | - | 4 | - | - | 3 | - |
| H/CF | - | - | - | - | - | - | - | - | 1 | 2 | 1 | - | - |
| H/COL | - | - | - | - | - | - | - | - | 5 | - | - | 1 | - |
| H/FS | - | - | - | - | - | - | - | - | 2 | - | - | 1 | - |
| M | 31 | 22 | - | 61 | - | - | - | - | - | - | - | - | - |
| MG | 2 | - | - | 3 | 2 | - | - | - | - | - | - | - | - |
| OTH | 2 | - | - | - | - | - | - | 1 | 2 | 1 | 1 | 2 | - |
| PZ/S | 1 | 1 | - | 4 | - | - | - | - | 1 | - | - | - | - |
| SD | - | - | - | - | - | - | - | - | 47 | - | 1 | 25 | 1 |
| *Total 1* | *376* | *25* | *2* | *227* | *152* | *2* | *3* | *111* | *98* | *3* | *4* | *39* | *1* |
| *Total 2* | *630* | | | | *268* | | | | *145* | | | | |

Quantities of Material From Blue Creek Categorised by Archaeological Context, Source, and Time Period

Table 3.32

| Context | Late Formative | | | | Early Classic | | | | Late Classic | | | | |
|---|---|---|---|---|---|---|---|---|---|---|---|---|---|
| | ELC | SMJ | IXT | *TTL%* | ELC | SMJ | IXT | *TTL%* | ELC | SMJ | IXT | MX | *TTL %* |
| C | 80.9 | | | *74.9* | 86.0 | 0.6 | 0.6 | *90.7* | 14.2 | | | | *14.5* |
| CF | 1.7 | 0.2 | | *3.0* | 4.5 | 0.6 | | *3.7* | 5.7 | | 0.9 | | *4.8* |
| COL | | | | | 0.6 | | | *0.4* | 0.9 | | | | *0.7* |
| FF | 0.2 | 0.2 | 0.2 | *0.6* | 4.5 | | 0.6 | *3.4* | 0.9 | | | | *0.7* |
| FS | 1.5 | | 0.2 | *1.3* | | | 0.6 | *0.4* | 12.3 | | | | *9.7* |
| H | | | | | | | | | 3.8 | | | | *4.8* |
| H/CF | | | | | | | | | 0.9 | 1.9 | 0.9 | | *2.8* |
| H/COL | | | | | | | | | 4.7 | | | | *4.1* |
| H/FS | | | | | | | | | 1.9 | | | | *2.1* |
| M | 7.7 | 5.5 | | *18.1* | | | | | | | | | |
| MG | 0.5 | | | *0.8* | 1.3 | | | *0.7* | | | | | |
| OTH | 0.5 | | | *0.3* | | | | *0.4* | 1.9 | 0.9 | 0.9 | | *4.1* |
| PZ/S | 0.2 | 0.2 | | *1.0* | | | | | 0.9 | | | | *0.7* |
| SD | | | | | | | | | 44.3 | | 0.9 | 0.9 | *51.0* |
| *Total 1* | *93.2* | *6.1* | *0.4* | *100* | *96.9* | *1.2* | *1.8* | *99.7* | *92.4* | *2.8* | *3.6* | *0.9* | *100.0* |
| *Total 2* | *99.7* | | | | *99.9* | | | | *99.7* | | | | |

Percentages of Material From Blue Creek Categorised by Archaeological Context, Source, and Time Period

Table 3.33

period. Ritual/Ceremonial 1 deposits were not represented at all in the Terminal Classic collection, while Ritual/Ceremonial 2 deposits were virtually identical to those from the preceding period, accounting for 23.9 per cent of the Terminal Classic collection versus 23.1 per cent of the Late Classic material. Material recovered from Workshop deposits increased from 4.3 per cent of the Late Classic collection to 15.2 per cent of the material from the Terminal Classic.

El Chayal and Ixtepeque material was recovered from Ritual/Ceremonial and Workshop contexts. San Martin Jilotepeque material was only recovered in limited quantities from Elite contexts. This latter material amounted to a scant 2.1 per cent of the Elite material.

Only the Central Peten and River areas reported material using this temporal designation (Tables 3.40 and 3.41, and Tables 3.42 to 3.45 respectively). The majority of material in the Central Peten area was from Indeterminate contexts (89.5 per cent), with the remaining 10.5 per cent being recovered from Non-Elite 1 contexts (Tables 3.40 and 3.41). The River area also yielded a large quantity of Indeterminate artefacts (47.6 per cent). The second largest collection from this area came from Workshop deposits which constituted 40.5 per cent of the Terminal Classic River collection. Elite 1 deposits contained the remaining 11.9 per cent of the collection (Tables 3.44 and 3.45).

### Summary of Social Contexts

Analysis of obsidian distribution by social contexts reveals a pattern of declining Ritual/Ceremonial 1 contexts in favour of Elite consumption. Starting in the Early Classic and continuing through the Terminal Classic Period Elite contexts account for the largest collections of obsidian. Ritual/Ceremonial 1 contexts are highest in the Late Formative then decline steadily through the Early and Late Classic periods, and disappear completely in the Terminal Classic. A shift in Ritual/Ceremonial activity occurs in the Late Classic Period when activities identified as Ritual/Ceremonial 2 are initiated. However, these deposits never exceed the quantities of material utilised in Elite contexts.

As with the data from the archaeological section, contextual information regarding the social distribution of material from different sources, revealed that El Chayal appeared in every context while San Martin Jilotepeque obsidian was restricted to specific deposits. The El Chayal material was the dominate, if not exclusive, source utilised for Ritual/Ceremonial contexts from the Late Formative until the Late Classic. During these periods El Chayal material consists of 100 to 93 per cent of the material. This is considerably higher than the average temporal distribution patterns which put El Chayal material at 71.9 per cent of the Late Formative collection, 83 per cent of the Early Classic collection and 73.2 per cent of the Late Classic collection (Table 3.1).

It is in the Terminal Classic period that use of El Chayal in Ritual/Ceremonial deposits drops below 90 per cent, accounting for only 63.3 per cent of the material in these deposits. This is still higher than the overall average for El Chayal material which was reported as comprising only 51 per cent of the Terminal Classic collection (Table 3.1). In the Terminal Classic Ritual/Ceremonial deposits the remaining 36.7 per cent consisted entirely of objects manufactured from Ixtepeque material. This increase in Ixtepeque material is consistent with the overall temporal distribution pattern previously documented (Table 3.1).

Objects manufactured from San Martin Jilotepeque appear primarily in Non-Elite contexts in the Late Formative through to the Late Classic. The Early Classic period when it is possible to detect greater social differentiation, the only context of determinable social standing in which this material appears is Non-Elite 1. It is likely that this material may have be utilised by Non-Elite 2 individuals however the collection from this context is too small to make conclusive observations. In the Late Classic period San Martin Jilotepeque material was also utilised by Elite individuals, albeit in a reduced capacity than in Non-Elite contexts. San Martin Jilotepeque material consists of 5.2 per cent of the Non-Elite deposits from this period and only 1.5 per cent of the Elite collection. In the Terminal Classic this amount increases to 2.1 per cent of the Elite collection. Non-Elite deposits from this last period are very small, making it difficult to discern a true pattern for the distribution.

### Summary of Contextual Analyses

Analysis of material by functional and archaeological contexts revealed a similar pattern of obsidian source utilisation. Functional analysis of the material disclosed that while El Chayal material appeared in all contexts it was predominately utilised in ritual contexts. It was also revealed that with a one piece exception, San Martin Jilotepeque material was not used in ritual contexts. When the ritual contexts were examined in closer detail by analysing the archaeological context of the material it was found that the aberrant piece of San Martin Jilotepeque material was contained in a cache, and that San Martin Jilotepeque obsidian was absent from all Tomb Shaft, Special Deposits and Mortuary Good collections. Where this material concentrated was in waste deposits, primarily construction fill and midden contexts, with smaller amounts being found in domestic contexts such as floor fill and surfaces.

Contrary to long-held standard beliefs that relate the distance, or exoticness, of an object with its value and therefore its role as status items in society, Mexican obsidian, which originates much further away than Guatemalan material, was not reserved for ritual or important activities. In the Early Classic Period this material was found in all identifiable functional contexts with the largest concentration being in waste deposits. During the Late Classic Mexican obsidian was reported only from ritual and waste contexts with the largest concentration being in the latter category. Archaeologically these contexts represent a concentration in construction fill deposits with smaller amounts being spread equally between cache, floor surface, floor fill and midden deposits. In the Late Classic the largest quantity was still found in construction fill with smaller, and again equal amounts, being recovered from humus and special deposits.

Analysis of obsidian distribution by social contexts reveals a pattern hitherto undisclosed by the previous functional and archaeological analyses. In the analysis of the material by functional contexts ritual activity is recorded as consuming 58.0 per cent of the Late Formative collection and was at it highest in the Early Classic where it utilised a total of 60.7 per cent of the obsidian from this collection. It also revealed that the decline in ritual activity in the Late Classic was gradual, declining to 36.2 per cent of the Late Classic collection before disappearing in the Terminal Classic. This pattern is borne out in an examination of the material by archaeological contexts which reveals large quantities of material in caches during the Late Formative and Early Classic Periods, shifting to Special Deposits in the Late Classic. Examination of the social contexts reveals a decline in Ritual/Ceremonial 1 contexts during the Early Classic and a rise in Elite consumption during this same period. This suggests that while caches was still an important and practised activity, the focus of this behaviour was shifting away from public venues into private, predominately elite loci. During the Late Formative ritual activity in public locations, the form of Ritual/Ceremonial 2 contexts, reveals both a slight increase in public activity (rising to 23.3 per cent from 18.4 per cent in the Early Classic) and a shift in the inherent behaviour of the event. In the Late Formative and Early Classic periods the primary pattern of ritual behaviour was caching, placing a collection of material in a single, concentrated deposit, secreted in a secure environment such as a stairblock, or beneath a structural modification. During the Late Classic these deposits changed so that they were no longer concentrated but often spread across an area (Driver 1999 :26; Garber 1981). This suggests that ritual activity at such locations was fulfilling a different function than in previous periods. A full discussion of such behavioural practices is beyond the scope of this analysis and warrants its own work, hence it shall only be noted in passing here as it affects the distribution pattern and consumption of obsidian, the focus of this work.

The above data shows clearly that source utilisation was tied to contexts, functional, archaeological and social. With El Chayal being the preferred material for ritual and mortuary activities and by elites, where it composes over 90 per cent of these deposits from the Late Formative to the Early Classic. Quantities greatly in excess of the period averages (Tables 3.1, 3.6, 3.7, 3.20, and 3.21). San Martin Jilotepeque material was confined primarily to waste deposits with small amounts appearing in domestic contexts (Tables 3.6 and 3.7). Ixtepeque and Mexican obsidian both appear to move between contexts, being found in similar deposits to El Chayal, but in distributions closer to those of San Martin Jilotepeque obsidian.

While the largest concentrations of El Chayal material are in ritual deposits (Tables 3.6 and 3.7) and caches (Tables 3.20 and 3.21), as a 'preferred material' this source appears in small amounts in virtually all contexts: this may been seen as a 'trickle down' effect, where the material filters down from the highest levels, and those contexts that have 'first pick' of the material, to other, lower status contexts. While El Chayal may be seen to 'trickle down' through the contexts, San Martin Jilotepeque rarely, if ever, 'trickles up'. An analogy would be those who could afford gold would disdain to use brass, while those who normally can afford only brass would covet and try to acquire what pieces of gold they could.

Now that we have established that source utilisation is linked to contextual deposits, and that specific materials were preferred for certain deposits, the next set of questions that arise from these observations are *'why was El Chayal preferred'*, and *'why was San Martin Jilotepeque resticted to waste and domestic deposits'*. The following section will investigate this question in relation to relative quality of the different obsidian sources.

| Ctxt | Middle Formative | | | | Late Formative | | | | Early Classic | | | | | EC/ LC | Late Classic | | | | | LC/TC | | Terminal Classic | | | |
|---|---|---|---|---|---|---|---|---|---|---|---|---|---|---|---|---|---|---|---|---|---|---|---|---|---|
| | EC | SM | IX | OT | EC | SM | IX | OT | EC | SM | IX | MX | OT | MX | EC | SM | IX | MX | OT | EC | IX | EC | SM | IX | OT |
| E1 | - | - | - | - | - | - | - | - | 138 | 1 | 1 | 1 | - | - | 53 | 2 | 17 | - | - | - | - | 3 | 1 | 1 | - |
| E1/2 | - | - | - | - | - | - | - | - | - | - | - | - | - | - | 43 | - | 17 | 1 | - | 26 | 16 | - | - | - | - |
| E2 | - | - | - | - | - | - | - | - | - | - | - | - | - | - | - | - | - | - | - | - | - | - | - | - | - |
| NE1 | - | - | - | - | 36 | 32 | - | - | 13 | 2 | 1 | 1 | - | - | 44 | 3 | 11 | - | - | - | - | - | - | 1 | - |
| NE2 | - | - | - | - | 7 | 1 | 1 | - | 1 | - | - | - | - | - | - | - | - | - | - | - | - | - | - | - | - |
| NE3 | - | - | - | - | - | - | - | - | - | - | - | - | - | - | - | - | - | - | - | - | - | - | - | - | - |
| RC1 | - | - | - | - | 325 | - | - | - | 55 | - | - | 3 | 1 | - | 1 | - | - | - | - | - | - | - | - | - | - |
| RC2 | - | - | - | - | - | - | - | - | - | - | - | - | - | - | 91 | - | 1 | - | - | 21 | 12 | - | - | - | - |
| WS | - | - | - | - | - | - | - | - | - | - | - | - | - | - | 9 | - | 8 | - | - | 4 | - | 5 | - | 12 | - |
| IND | 12 | 48 | 5 | 2 | 77 | 34 | 36 | 13 | 62 | 6 | 10 | 18 | 6 | 3 | 53 | 13 | 27 | 1 | 3 | - | - | 18 | 1 | 16 | 1 |
| *TTL1* | 12 | 48 | 5 | 2 | *445* | *67* | *37* | *13* | *269* | *9* | *12* | *23* | *7* | *3* | *294* | *18* | *81* | *2* | *3* | *51* | *28* | *26* | *2* | *30* | *1* |
| *TTL 2* | *67* | | | | *562* | | | | *320* | | | | | *3* | *398* | | | | | *79* | | *59* | | | |

Quantities of Material by Social Context, Source and Time Period

Table 3. 34

| Ctxt | Middle Formative | | | | Late Formative | | | | Early Classic | | | | | EC/ LC | Late Classic | | | | | LC/TC | | Terminal Classic | | | |
|---|---|---|---|---|---|---|---|---|---|---|---|---|---|---|---|---|---|---|---|---|---|---|---|---|---|
| | EC | SM | IX | OT | EC | SM | IX | OT | EC | SM | IX | MX | OT | MX | EC | SM | IX | MX | OT | EC | IX | EC | SM | IX | OT |
| E1 | - | - | - | - | - | - | - | - | 43.1 | 0.3 | 0.3 | 0.3 | - | - | 13.3 | 0.5 | 4.3 | - | - | - | - | 5.1 | 1.7 | 1.7 | - |
| E2 | - | - | - | - | - | - | - | - | - | - | - | - | - | - | 10.8 | - | 4.3 | 0.2 | - | 32.9 | 20.3 | - | - | - | - |
| NE1 | - | - | - | - | 6.4 | 5.7 | - | - | 4.1 | 0.6 | 0.3 | 0.3 | - | - | 11.0 | 0.8 | 2.8 | - | - | - | - | - | - | 1.7 | - |
| NE2 | - | - | - | - | 1.2 | 0.2 | 0.2 | - | 0.3 | - | - | - | - | - | - | - | - | - | - | - | - | - | - | - | - |
| NE3 | - | - | - | - | - | - | - | - | - | - | - | - | - | - | - | - | - | - | - | - | - | - | - | - | - |
| RC1 | - | - | - | - | 57.8 | - | - | - | 17.2 | - | - | 0.9 | 0.3 | - | 0.2 | - | - | - | - | - | - | - | - | - | - |
| RC2 | - | - | - | - | - | - | - | - | - | - | - | - | - | - | 22.9 | - | 0.2 | - | - | 26.6 | 15.2 | - | - | - | - |
| WS | - | - | - | - | - | - | - | - | - | - | - | - | - | - | 2.3 | - | 2.0 | - | - | 5.1 | - | 8.5 | - | 20.3 | - |
| IND | 17.9 | 71.6 | 7.5 | 3.0 | 13.7 | 6.0 | 6.4 | 2.3 | 19.4 | 1.9 | 3.1 | 5.6 | 1.9 | 100 | 13.3 | 3.3 | 6.8 | 0.2 | 0.7 | - | - | 30.5 | 1.7 | 27.1 | 1.7 |
| *TTL1* | 17.9 | 71.6 | 7.5 | 3.0 | *79.1* | *11.9* | *6.6* | *2.3* | *84.1* | *2.8* | *3.7* | *7.1* | *2.2* | *100* | *73.8* | *4.6* | *20.4* | *0.4* | *0.7* | *64.6* | *35.3* | *44.1* | *3.4* | *50.8* | *1.7* |
| *TTL 2* | *100.0* | | | | *99.9* | | | | *99.9* | | | | | *100* | *99.9* | | | | | *100.1* | | *100.0* | | | |

Percentages of Material from All Sites by Social Context, Source and Time Period

Table 3.35a

| Context | Late Formative | | | Early Classic | | | Late Classic | | | Terminal Classic | | |
|---|---|---|---|---|---|---|---|---|---|---|---|---|
| | ELC | SMJ | IXT | ELC | SMJ | IXT | ELC | SMJ | IXT | ELC | SMJ | IXT |
| E1 | - | - | - | - | - | - | 17 | - | 3 | - | - | - |
| NE1 | - | - | - | 12 | - | 1 | - | - | - | - | - | - |
| IND | 16 | - | - | 1 | - | 1 | - | - | - | 1 | - | - |
| *Total 1* | *16* | - | - | *13* | - | *2* | *17* | - | *3* | *1* | - | - |
| *Total 2* | *16* | | | *15* | | | *20* | | | *1* | | |

Quantities of Coastal Material by Social Context, Source and Time Period

Table 3.36

| Context | Late Formative | | | Early Classic | | | Late Classic | | | Terminal Classic | | |
|---|---|---|---|---|---|---|---|---|---|---|---|---|
| | ELC | SMJ | IXT | ELC | SMJ | IXT | ELC | SMJ | IXT | ELC | SMJ | IXT |
| E1 | - | - | - | - | - | - | 85.0 | - | 15.0 | - | - | - |
| NE1 | - | - | - | 80.0 | - | 6.7 | - | - | - | - | - | - |
| IND | 100.0 | - | - | 6.7 | - | 6.7 | - | - | - | 100.0 | - | - |
| *Total 1* | *100.0* | - | - | *86.7* | - | *13.4* | *85.0* | - | *15.0* | *100.0* | - | - |
| *Total 2* | *100.0* | | | *100.1* | | | *100.0* | | | *100.0* | | |

Percentages Coastal Material by Social Context, Source and Time Period

Table3.37

| Context | Middle Formative | | | Late Formative | | | Early Classic | | | | Late Classic | | |
|---|---|---|---|---|---|---|---|---|---|---|---|---|---|
| | ELC | SMJ | IXT | ELC | SMJ | IXT | ELC | SMJ | IXT | MX | ELC | SMJ | IXT |
| RC1 | - | - | - | - | - | - | 51 | - | - | 3 | - | - | - |
| RC2 | - | - | - | - | - | - | - | - | - | - | 86 | - | - |
| IND | 1 | 1 | 2 | 1 | - | 1 | 18 | - | 2 | 2 | 1 | - | - |
| *Total 1* | *1* | *1* | *2* | *1* | - | *1* | *69* | - | *2* | *5* | *87* | - | - |
| *Total 2* | *4* | | | *2* | | | *76* | | | | *87* | | |

Quantities of Material from Inland Sites by Social Context, Source and Time Period

Table 3.38

| Context | Middle Formative | | | Late Formative | | | Early Classic | | | | Late Classic | | |
|---|---|---|---|---|---|---|---|---|---|---|---|---|---|
| | ELC | SMJ | IXT | ELC | SMJ | IXT | ELC | SMJ | IXT | MX | ELC | SMJ | IXT |
| RC1 | - | - | - | - | - | - | 67.1 | - | - | 3.9 | - | - | - |
| RC2 | - | - | - | - | - | - | - | - | - | - | 98.1 | - | - |
| IND | 25 | 25 | 50 | 50 | - | 50 | 23.7 | - | 2.6 | 2.6 | 1.1 | - | - |
| *Total 1* | *25* | *25* | *50* | *50* | - | *50* | *90.8* | - | - | - | *99.9* | - | - |
| *Total 2* | *100* | | | *100* | | | *99.9* | | | | *99.9* | | |

Percentages of Material from Inland Sites by Social Context, Source and Time Period

Table 3.39

| Context | Middle Formative | | | | Late Formative | | | Early Classic | | | | EC/ LC | Late Classic | | | | | Terminal Classic | | | |
|---|---|---|---|---|---|---|---|---|---|---|---|---|---|---|---|---|---|---|---|---|---|
| | EC | SM | IX | OT | EC | SM | IX | EC | SM | MX | OT | MX | EC | SM | IX | MX | OT | EC | SM | IX | OT |
| E1 | - | - | - | - | - | - | - | - | - | 1 | - | - | - | - | - | - | - | - | - | - | - |
| E1/2 | - | - | - | - | - | - | - | - | - | - | - | - | - | - | - | - | - | - | - | - | - |
| NE1 | - | - | - | - | 3 | 10 | - | 1 | 2 | 1 | - | - | 34 | 3 | - | - | - | - | - | 1 | - |
| RC1 | - | - | - | - | - | - | - | 3 | - | - | - | - | 1 | - | - | - | - | - | - | - | - |
| RC2 | - | - | - | - | - | - | - | - | - | - | - | - | 5 | - | - | - | - | - | - | - | - |
| WS | - | - | - | - | - | - | - | - | - | - | - | - | - | - | - | - | - | - | - | - | - |
| IND | 11 | 47 | 3 | 2 | 32 | 31 | 4 | 21 | 6 | 16 | 3 | 3 | 44 | 13 | 4 | 1 | 3 | 13 | 1 | - | 1 |
| *Total 1* | *11* | *47* | *3* | *2* | *35* | *41* | *4* | *25* | *8* | *18* | *3* | *3* | *84* | *16* | *4* | *1* | *3* | *13* | *1* | *2* | *3* |
| *Total 2* | *63* | | | | *80* | | | *54* | | | | *3* | *108* | | | | | *19* | | | |

Quantities of Material from Peten Sites by Social Context, Source and Time Period

Table 3.40

| Context | Middle Formative | | | | Late Formative | | | Early Classic | | | | EC/ LC | Late Classic | | | | | Terminal Classic | | | |
|---|---|---|---|---|---|---|---|---|---|---|---|---|---|---|---|---|---|---|---|---|---|
| | EC | SM | IX | OT | EC | SM | IX | EC | SM | MX | OT | MX | EC | SM | IX | MX | OT | EC | SM | IX | OT |
| E1 | | | | | | | | | | 1.9 | | | | | | | | | | | |
| E1/2 | | | | | | | | | | | | | | | | | | | | | |
| NE1 | | | | | 3.8 | 12.5 | | 1.9 | 3.7 | 1.9 | | | 31.5 | 2.8 | | | | | | 10.5 | |
| RC1 | | | | | | | | 5.5 | | | | | 0.9 | | | | | | | | |
| RC2 | | | | | | | | | | | | | 4.6 | | | | | | | | |
| WS | | | | | | | | | | | | | | | | | | | | | |
| IND | 17.5 | 74.6 | 4.8 | 3.2 | 40.0 | 38.8 | 5.0 | 38.9 | 11.1 | 29.6 | 5.5 | 100 | 40.7 | 12.0 | 3.7 | 0.9 | 2.8 | 68.4 | 5.3 | 10.5 | 15.8 |
| *Total 1* | 17.5 | 74.6 | 4.8 | 3.2 | *43.8* | *51.3* | *5.0* | *46.3* | *14.8* | *33.4* | *5.5* | *100* | *77.7* | *14.8* | *3.7* | *0.9* | *2.8* | *68.4* | *5.3* | *10.5* | *15.8* |
| *Total 2* | *100.1* | | | | *100.1* | | | *100.0* | | | | *100* | *99.9* | | | | | *100.0* | | | |

Percentages of Material from Peten Sites by Social Context, Source and Time Period

Table 3.41

| Context | Late Formative | | | | Early Classic | | | Late Classic | | | Late/Terminal Classic | | Terminal Classic | | |
|---|---|---|---|---|---|---|---|---|---|---|---|---|---|---|---|
| | ELC | SMJ | IXT | OTH | ELC | IXT | OTH | ELC | SMJ | IXT | ELC | IXT | ELC | SMJ | IXT |
| E1 | - | - | - | - | - | - | - | 5 | 1 | 14 | - | - | 3 | 1 | 1 |
| E2 | - | - | - | - | - | - | - | 3 | - | 17 | 26 | 16 | - | - | - |
| NE2 | - | - | - | - | - | - | - | 10 | - | 11 | - | - | - | - | - |
| RC2 | - | - | - | - | - | - | - | - | - | - | 21 | 12 | - | - | - |
| WS | - | - | - | - | - | - | - | 9 | - | 8 | 4 | - | 5 | - | 12 |
| IND | 17 | 1 | 30 | 13 | 10 | 6 | 1 | 8 | - | 20 | - | - | 4 | - | 16 |
| *Total 1* | *17* | *1* | *30* | *13* | *10* | *6* | *1* | *35* | *1* | *70* | *51* | *28* | *12* | *1* | *29* |
| *Total 2* | *61* | | | | *17* | | | *106* | | | *79* | | *42* | | |

Quantities of Material from River Sites, Excluding Blue Creek, by Social Context, Source and Time Period

Table 3.42

| Context | Late Formative | | | | Early Classic | | | Late Classic | | | Late/Terminal Classic | | Terminal Classic | | |
|---|---|---|---|---|---|---|---|---|---|---|---|---|---|---|---|
| | ELC | SMJ | IXT | OTH | ELC | IXT | OTH | ELC | SMJ | IXT | ELC | IXT | ELC | SMJ | IXT |
| E1 | - | - | - | - | - | - | - | 4.7 | 0.9 | 13.2 | - | - | 7.1 | 2.4 | 2.4 |
| E2 | - | - | - | - | - | - | - | 2.8 | - | 16.0 | 32.9 | 20.3 | - | - | - |
| NE2 | - | - | - | - | - | - | - | 9.4 | - | 10.4 | - | - | - | - | - |
| RC2 | - | - | - | - | - | - | - | - | - | - | 26.6 | 15.2 | - | - | - |
| WS | - | - | - | - | - | - | - | 8.5 | - | 7.5 | 5.1 | - | 11.9 | - | 28.6 |
| IND | 27.9 | 1.6 | 49.2 | 21.3 | 58.8 | 35.3 | 5.9 | 7.5 | - | 18.9 | - | - | 9.5 | - | 38.1 |
| *Total 1* | *27.9* | *1.6* | *49.2* | *21.3* | *58.8* | *35.3* | *5.9* | *32.9* | *0.9* | *66.0* | *64.6* | *35.5* | *28.5* | *2.4* | *69.1* |
| *Total 2* | *100.0* | | | | *100.0* | | | *99.8* | | | *100.1* | | *100.0* | | |

Percentages of Material from River Sites, Excluding Blue Creek, by Social Context, Source and Time Period

Table 3.43

| Context | Late Formative | | | | Early Classic | | | Late Classic | | | | Late/Terminal Classic | | Terminal Classic | | |
|---|---|---|---|---|---|---|---|---|---|---|---|---|---|---|---|---|
| | ELC | SMJ | IXT | OTH | ELC | IXT | OTH | ELC | SMJ | IXT | MX | ELC | IXT | ELC | SMJ | IXT |
| E1 | - | - | - | - | 138 | 1 | - | 36 | 2 | 14 | - | - | - | 3 | 1 | 1 |
| E2 | - | - | - | - | - | - | - | 43 | - | 17 | 1 | 26 | 16 | - | - | - |
| NE1 | 33 | 22 | - | - | - | - | - | - | - | - | - | | | | | |
| NE2 | 7 | 1 | 1 | - | 1 | - | - | 10 | - | 11 | - | - | - | - | - | - |
| RC1 | 325 | - | - | - | 1 | - | 1 | - | - | - | - | | | | | |
| RC2 | - | - | - | - | - | - | - | - | - | 1 | - | 21 | 12 | - | - | - |
| WS | - | - | - | - | - | - | - | 9 | - | 8 | - | 4 | - | 5 | - | 12 |
| IND | 28 | 3 | 31 | 13 | 22 | 7 | 3 | 8 | - | 23 | - | - | - | 4 | - | 16 |
| *Total 1* | *393* | *26* | *32* | *13* | *162* | *8* | *4* | *106* | *2* | *74* | *1* | *51* | *28* | *12* | *1* | *29* |
| *Total 2* | *464* | | | | *174* | | | *183* | | | | *79* | | *42* | | |

Quantities of Material from River Sites by Social Context, Source and Time Period

Table 3.44

| Context | Late Formative | | | | Early Classic | | | Late Classic | | | | Late/Terminal Classic | | Terminal Classic | | |
|---|---|---|---|---|---|---|---|---|---|---|---|---|---|---|---|---|
| | ELC | SMJ | IXT | OTH | ELC | IXT | OTH | ELC | SMJ | IXT | MX | ELC | IXT | ELC | SMJ | IXT |
| E1 | | | | | 79.3 | 0.6 | | 19.7 | 1.1 | 7.6 | | | | 7.1 | 2.4 | 2.4 |
| E2 | | | | | | | | 23.5 | | 9.3 | 0.5 | 32.9 | 20.3 | | | |
| NE1 | 7.1 | 4.7 | | | | | | | | | | | | | | |
| NE2 | 1.5 | 0.2 | 0.2 | | 0.6 | | | 5.5 | | 6.0 | | | | | | |
| RC1 | 70.0 | | | | 0.6 | | 0.6 | | | | | | | | | |
| RC2 | | | | | | | | | | 0.5 | | 26.6 | 15.2 | | | |
| WS | | | | | | | | 4.9 | | 4.4 | | 5.1 | | 11.9 | - | 28.6 |
| IND | 6.0 | 0.6 | 6.7 | 2.8 | 12.6 | 4.0 | 1.7 | 4.4 | | 12.6 | | | | 9.5 | - | 38.1 |
| *Total 1* | *84.6* | *5.5* | *6.9* | *2.8* | *93.1* | *4.6* | *2.3* | *58.0* | *1.1* | *40.4* | *0.5* | *64.6* | *35.5* | *28.5* | *2.4* | *69.1* |
| *Total 2* | *99.8* | | | | *100.0* | | | *100.0* | | | | *100.1* | | *100.0* | | |

Percentages of Material from River Sites by Social Context, Source and Time Period

Table 3.45

| Context | Late Formative | | | | Early Classic | | | | Late Classic | | | | |
|---|---|---|---|---|---|---|---|---|---|---|---|---|---|
| | ELC | SMJ | IXT | NA | ELC | SMJ | IXT | NA | ELC | SMJ | IXT | NA | MX |
| E1 | - | - | - | - | 138 | 1 | - | 104 | 31 | 1 | - | 6 | - |
| E2 | - | - | - | - | - | - | - | - | 40 | - | - | 22 | 1 |
| NE1 | 33 | 22 | - | 61 | - | - | - | - | - | - | - | - | - |
| NE2 | 7 | 1 | 1 | 11 | 1 | - | - | - | 1 | - | - | 1 | - |
| NE3 | - | - | - | 2 | - | - | - | - | - | - | - | - | - |
| RC1 | 325 | - | - | 145 | 1 | - | 1 | 3 | - | - | - | - | - |
| RC2 | - | - | - | - | - | - | - | - | 6 | - | 1 | 5 | - |
| IND | 11 | 2 | 1 | 8 | 12 | 1 | 2 | 4 | 20 | 2 | 3 | 5 | - |
| *Total 1* | *376* | *25* | *2* | *227* | *152* | *1* | *3* | *111* | *98* | *3* | *4* | *39* | *1* |
| *Total 2* | *630* | | | | *268* | | | | *145* | | | | |

Quantities of Material From Blue Creek by Social Context, Source and Time Period

Table 3.46

| Context | Late Formative | | | Early Classic | | | Late Classic | | | |
|---|---|---|---|---|---|---|---|---|---|---|
| | ELC | SMJ | IXT | ELC | SMJ | IXT | ELC | SMJ | IXT | MX |
| E1 | | | | 87.9 | 0.6 | | 29.2 | 0.9 | | |
| E2 | | | | | | | 37.7 | | | 0.9 |
| NE1 | 8.2 | 5.5 | | | | | | | | |
| NE2 | 1.7 | 0.2 | 0.2 | 0.6 | | | 0.9 | | | |
| NE3 | | | | | | | | | | |
| RC1 | 80.6 | | | 0.6 | | 0.6 | | | | |
| RC2 | | | | | | | 5.7 | | 0.9 | |
| IND | 2.7 | 0.5 | 0.2 | 7.6 | 0.6 | 1.3 | 18.9 | 1.9 | 2.8 | |
| *Total 1* | *93.2* | *6.2* | *0.4* | *96.7* | *1.2* | *1.9* | *92.4* | *2.8* | *3.7* | *0.9* |
| *Total 2* | *99.8* | | | *99.8* | | | *99.8* | | | |

Percentages of Material From Blue Creek by Social Context, Source and Time Period

Table 3.47

## SECTION 3
## QUALITY, CONTEXT AND TIME MODEL

### INTRODUCTION

This section will address questions arising from the previous section's conclusion that certain obsidian sources are utilised preferentially for specific activities. Specifically it will endeavour to determine *'why El Chayal obsidian was preferred'*, and *'why San Martin Jilotepeque was restricted primarily to waste and domestic deposits'*.

Crabtree, in his work on obsidian blade production, noted that the fewer inclusions present in the obsidian resulted in better manufactured blades. Therefore the simplest query is *'do obsidian sources vary in quality'*, quality being defined here as different gradients and quantities of inclusions. A full explanation of the technique used may be found in Chapter 2, Section 1 under the heading 'Questions of Quality'. Suffice it to say here that obsidian artefacts were examined for the presence and types of inclusions, then classified into three 'grades' with Q-1 representing the finest material, and Q-3 containing the largest number and types of inclusions. As shown in Figure 2.1, it is possible to distinguish potential grades of obsidian, and, as demonstrated in Tables 3.48 and 3.49, the percentages of each quality vary between sources.

As the only material with a broad spectrum of contexts and temporal periods available for examination was the Blue Creek collection, the following discussions on quality are based on these artefacts. It should be noted, that as this analysis was conducted after several of the pieces had been sourced by NAA, some of the artefacts were no longer available for visual examination. Hence, the quantities presented in the quality analysis below differ from those used in the previous sections in discussions of obsidian source

### QUALITIES OF OBSIDIAN

Ixtepeque obsidian had the largest percentage of high quality (Q-1) material (88.2 per cent) (Table 3.48, 3.49). As this material is visually identifiable by its clarity and transparent brown colour, the high percentage of Q-1 material was expected. El Chayal obsidian also possessed a high percentage of Q-1 material (68.3 per cent), with percentages of the other qualities declining as one descends in quality. Half of the San Martin Jilotepeque material (50.0 per cent) is middle quality (Q-2), with the second largest percentage being low quality (Q-3) material (38.9 per cent). The San Martin Jilotepeque source possess the lowest proportion of Q-1 material (11.1 per cent).

During the Late Formative Period, Q-1 material accounts for the bulk of the obsidian collection with 59.2 per cent (Table 3.52). Although Q-1 material remains dominant in the Early Classic, it is reduced slightly to 50.6 per cent of the collection. Quality 1 material continues to decrease in the Late Classic period where, at 43.4 per cent, it is supplanted by Q-2 as the dominant material. During the Late Classic, Q-2 material accounted for 49.3 per cent of the collection.

| | Q-1 | Q-2 | Q-3 | Total |
|---|---|---|---|---|
| El Chayal | 449 | 197 | 11 | 657 |
| San Martin Jilotepeque | 2 | 9 | 7 | 18 |
| Ixtepeque | 15 | 2 | - | 17 |
| Zaragoza | 1 | - | - | 1 |

Quanties of Each Quality per Obsidian Source
(includes material with no temporal context)

Table 3.49

| | Q-1 | Q-2 | Q-3 | Total |
|---|---|---|---|---|
| El Chayal | 68.3 | 30.0 | 1.7 | 100.0 |
| San Martin Jilotepeque | 11.1 | 50.0 | 38.9 | 100.0 |
| Ixtepeque | 88.2 | 11.8 | - | 100.0 |
| Zaragoza | 100.0 | - | - | 100.0 |

Percentages of Each Quality per Obsidian Source
(includes material with no temporal context)

Table 3.49

| DATE | Q-1 | Q-2 | Q-3 | TOTAL |
|---|---|---|---|---|
| LF | 351 | 198 | 44 | 593 |
| EC | 133 | 108 | 22 | 263 |
| LC | 59 | 67 | 10 | 136 |
| ND | 39 | 24 | 11 | 74 |
| TOTAL | 582 | 397 | 87 | 1066 |

Quantities of Each Quality of Obsidian by Time Period

Table 3.50

| DATE | Q-1 | Q-2 | Q-3 | TOTAL |
|---|---|---|---|---|
| LF | 59.2 | 33.4 | 7.4 | 100 |
| EC | 50.6 | 41.1 | 8.3 | 100 |
| LC | 43.4 | 49.3 | 7.3 | 100 |
| ND | 52.7 | 32.4 | 14.9 | 100 |

Percentages of Each Quality of Obsidian by Time Period

Table 3.51

### Obsidian Quality By Functional And Archaeological Contexts

Since it is possible to distinguish different qualities of obsidian, the next question needs be *'does the quality of obsidian relate to the contextual distribution of the material?'*. It is natural to assume that finer quality material would appear in higher status contexts, and that the reverse would also be true, that lower quality material would predominate in lower status context.

Following this premise, one might assume that the previous association between El Chayal and status deposits is directly connected to the fact that this material is generally of high quality, and that this Q-1 material is utilised in the ritual contexts. However, an analysis of the material by functional context and quality reveals that while the largest proportion of ritual material is *generally* composed of Q-1 pieces, it is not *exclusively* composed of material, and items manufactured from all qualities appear in ritual contexts (Table 3.52 and 3.53).

Conversely, knowing that San Martin Jilotepeque material is primarily associated with waste contexts, and that the majority of this material is of medium to low grade, it would follow that the natural assumption would be that waste contexts were composed primarily of Q-2 to Q-3 objects. However, this is not the case. As the evidence in Tables 3.52 and 3.53 demonstrates these deposits are often higher in Quality 1 material.

Analysis of the material by functional context also reveals that the Q-1 material was not concentrated in the ritual and burial contexts, but in all contexts, in a manner similar to the El Chayal material (Tables 3.6 and 3.7). A closer examination of the distribution pattern of the different qualities of obsidian combined with the previously noted pattern of a virtual exclusivity of El Chayal material in cache contexts reveals that, while Q-1 material predominates, El Chayal material of all qualities was included in caches during the Late Formative and Early Classic Period (Table 3.54). In the Late Classic, when caching has been replaced as the predominant ritual activity by special deposits, we continue to see a broad consumption pattern of El Chayal material. What little caching does occur in the Late Classic contains only Q-1, El Chayal obsidian, while the special deposits are primarily of Q-2, El Chayal obsidian. This suggests that the preference for El Chayal material was not based on quality or manufacturing. It is possible that the preference for this material is based on the origin of the material, the manner in which is was traded or on nuances of economic interactions that are currently obscure.

### Obsidian Quality By Social Context

As noted in the discussion of quality and obsidian sources, the majority of El Chayal artefacts are of the finest grade material, and obsidian from San Martin Jilotepeque is composed primarily of second and third grade material. It would be easy therefore to assume that the majority of high status ritual or elite contexts are composed primarily, or exclusively, of high quality obsidian. However, as with the issue surrounding the distribution of the various qualities of obsidian in functional and archaeological contexts, this would be misleading.

| Context | Late Formative | | | Early Classic | | | Late Classic | | |
|---|---|---|---|---|---|---|---|---|---|
| | Q-1 | Q-2 | Q-3 | Q-1 | Q-2 | Q-3 | Q-1 | Q-2 | Q-3 |
| Ritual | 308 | 139 | 17 | 121 | 100 | 19 | 31 | 54 | 8 |
| Waste | 33 | 49 | 24 | 6 | 4 | 1 | 17 | 5 | 1 |
| Domestic | 6 | 4 | 1 | 4 | 3 | 2 | 9 | 3 | 1 |
| Burial | 2 | 2 | 1 | 1 | 1 | - | - | - | - |
| Indeterminate | 2 | 4 | 1 | 1 | - | - | 2 | 5 | - |
| *Total 1* | *351* | *198* | *44* | *133* | *108* | *22* | *59* | *67* | *10* |
| *Total 2* | *593* | | | *263* | | | *136* | | |

Quanties of Blue Creek Material by Quality, Functional Context and Time Period

Table 3. 52

| Context | Late Formative | | | Early Classic | | | Late Classic | | |
|---|---|---|---|---|---|---|---|---|---|
| | Q-1 | Q-2 | Q-3 | Q-1 | Q-2 | Q-3 | Q-1 | Q-2 | Q-3 |
| Ritual | 52.0 | 23.4 | 2.9 | 46.0 | 38.0 | 7.2 | 22.8 | 39.7 | 5.9 |
| Waste | 5.6 | 8.3 | 4.0 | 2.3 | 1.5 | 0.4 | 12.5 | 3.7 | 0.7 |
| Domestic | 1.0 | 0.7 | 0.2 | 1.5 | 1.1 | 0.8 | 6.6 | 2.2 | 0.7 |
| Burial | 0.3 | 0.3 | 0.2 | 0.4 | 0.4 | - | - | - | - |
| Indeterminate | 0.3 | 0.7 | 0.2 | 0.4 | - | - | 1.5 | 3.7 | - |
| *Total 1* | *59.2* | *33.4* | *7.5* | *50.6* | *41.0* | *8.4* | *43.4* | *49.3* | *7.3* |
| *Total 2* | *100.1* | | | *100.0* | | | *100.0* | | |

Percentages of Blue Creek Material by Quality, Functional Context and Time Period

Table 3. 53

An examination of the social contexts of the different qualities of obsidian reveals that Q-1 material, while concentrated in Ritual/Ceremonial contexts during the Late Formative Period, is dispersed throughout the occupational contexts. Within these contexts, the largest proportion of the material is in Non-Elite 1 contexts and the quantities decline as the status of the deposits decreases. Of interest is that the largest proportion of Non-Elite 1 and Non-Elite 2 contexts is composed of Q-2 material.
In the Early Classic Period, this pattern shifts, with the majority of material being concentrated in the hands of the Elite 1 individuals. The largest proportion of Elite material is Q-1. However, as the Elite contexts utilise 92.8 per cent of the Early Classic obsidian, and as Q-1 material dominates these collections, the large quantity of Q-1 material is expected (Table 3.51).

Although the majority of Late Classic obsidian (70.3 per cent) is still concentrated in the hands of the Elite, the largest proportion of the material (44.2 per cent) is in the hands of Elite 2 individuals. However, the majority of the Elite 1 material is Q-1, while the majority of the Elite 2 material is Q-2. This suggests that while the Elite 2 individuals were consuming more obsidian than their peers inside the site core, they were not necessarily acquiring larger quantities of high grade material. As noted above, Q-2 material dominates this period, albeit by only a slight margin (Table 3.51). It is possible that the distribution pattern noted for the Elite 2 collection is an extension of this temporal pattern.

## SUMMARY

As demonstrated above, it is possible to distinguish different qualities, or grades, of obsidian. The proportions of these qualities vary between sources, with El Chayal and Ixtepeque sources possessing high quantities of high quality material. San Martin Jilotepeque obsidian consists primarily of middle, or second grade material, followed by third grade material.

The natural inclination of assuming that high quality items would appear in deposits of higher status (elite) or greater importance (ritual) has been demonstrated as erroneous. Obsidian artefacts appear in the various functional, archaeological, and social contexts without regard to quality. Ritual deposits are not composed exclusively of Q-1 material, and conversely waste deposits are not limited to Q-3 artefacts. However, as noted in Chapter 3, Section 2, ritual and burial deposits are composed almost exclusively of El Chayal material. It appears that material chosen for these deposits was not selected based on concepts of quality or considered for paucity of inclusions. This suggests that the choice of obsidian was based on source of origin, exchange networks, or social or political motives not clearly apparent. The following chapter will explore these possibilities in detail.

| Context | Source | LF | | | EC | | | LC | | |
|---|---|---|---|---|---|---|---|---|---|---|
| | | Q1 | Q2 | Q3 | Q1 | Q2 | Q3 | Q1 | Q2 | Q3 |
| C | ELC | 250 | 72 | 3 | 88 | 46 | 1 | 17 | - | - |
| | SMJ | - | - | - | - | - | 1 | - | - | - |
| | IXT | - | - | - | 1 | - | - | - | - | - |
| | **Total** | **250** | **72** | **3** | **89** | **46** | **2** | **17** | **-** | **-** |
| CF | ELC | 7 | - | - | 5 | 2 | - | 4 | 2 | 1 |
| | SMJ | - | 1 | - | - | - | 1 | - | - | - |
| | IXT | - | - | - | - | - | - | 1 | - | - |
| | **Total** | **7** | **1** | **-** | **5** | **2** | **1** | **5** | **2** | **1** |
| COL | ELC | - | - | - | - | 1 | - | 1 | - | - |
| | SMJ | - | - | - | - | - | - | - | - | - |
| | IXT | - | - | - | - | - | - | - | - | - |
| | **Total** | **-** | **-** | **-** | **-** | **1** | **-** | **1** | **-** | **-** |
| FF | ELC | - | - | - | 3 | 2 | 1 | - | 1 | - |
| | SMJ | - | - | 1 | - | - | - | - | - | - |
| | IXT | 1 | - | - | - | - | - | - | - | - |
| | **Total** | **1** | **-** | **1** | **3** | **2** | **1** | **-** | **1** | **-** |
| FS | ELC | 3 | 3 | - | - | - | - | 9 | 2 | 1 |
| | SMJ | - | - | - | - | - | - | - | - | - |
| | IXT | 1 | - | - | 1 | - | - | - | - | - |
| | **Total** | **4** | **3** | **-** | **1** | **-** | **-** | **9** | **2** | **1** |
| H | ELC | - | - | - | - | - | - | 4 | - | - |
| | SMJ | - | - | - | - | - | - | - | - | - |
| | IXT | - | - | - | - | - | - | - | - | - |
| | **Total** | **-** | **-** | **-** | **-** | **-** | **-** | **4** | **-** | **-** |
| H/MIX | ELC | - | - | - | - | - | - | 4 | 2 | - |
| | SMJ | - | - | - | - | - | - | 1 | - | - |
| | IXT | - | - | - | - | - | - | 1 | - | - |
| | **Total** | **-** | **-** | **-** | **-** | **-** | **-** | **6** | **2** | **-** |
| M | ELC | 18 | 5 | - | - | - | - | - | - | - |
| | SMJ | 1 | 5 | 3 | - | - | - | - | - | - |
| | IXT | - | - | - | - | - | - | - | - | - |
| | **Total** | **19** | **10** | **3** | **-** | **-** | **-** | **-** | **-** | **-** |
| MG | ELC | - | 2 | - | 1 | 1 | - | - | - | - |
| | SMJ | - | - | - | - | - | - | - | - | - |
| | IXT | - | - | - | - | - | - | - | - | - |
| | **Total** | **-** | **2** | **-** | **1** | **1** | **-** | **-** | **-** | **-** |
| OTH | ELC | 1 | 1 | - | - | - | - | 2 | - | - |
| | SMJ | - | - | - | - | - | - | - | 1 | - |
| | IXT | - | - | - | - | - | - | - | 1 | - |
| | **Total** | **1** | **1** | **-** | **-** | **-** | **-** | **2** | **2** | **-** |
| PZ/S | ELC | 1 | - | - | - | - | - | - | 1 | - |
| | SMJ | - | - | 1 | - | - | - | - | - | - |
| | IXT | - | - | - | - | - | - | - | - | - |
| | **Total** | **1** | **-** | **1** | **-** | **-** | **-** | **-** | **1** | **-** |
| SD | ELC | - | - | - | - | - | - | 18 | 27 | 2 |
| | SMJ | - | - | - | - | - | - | - | - | - |
| | IXT | - | - | - | - | - | - | - | 1 | - |
| | MX | - | - | - | - | - | - | 1 | - | - |
| | **Total** | **-** | **-** | **-** | **-** | **-** | **-** | **19** | **28** | **2** |

Quantities of Material by Archaeological Context, Source and Quality

Table 3.54

| Context | Late Formative | | | Early Classic | | | Late Classic | | |
|---|---|---|---|---|---|---|---|---|---|
| | Q-1 | Q-2 | Q-3 | Q-1 | Q-2 | Q-3 | Q-1 | Q-2 | Q-3 |
| E1 | - | - | - | 123 | 102 | 19 | 18 | 17 | 1 |
| E2 | - | - | - | | - | - | 20 | 37 | 4 |
| NE1 | 24 | 48 | 21 | 1 | - | - | - | 1 | - |
| NE2 | 6 | 8 | 4 | | - | - | - | - | - |
| NE3 | 1 | - | 1 | | - | - | - | - | - |
| RC1 | 307 | 138 | 17 | | 1 | - | - | - | - |
| RC2 | - | - | - | | - | - | - | - | - |
| IND | 13 | 4 | 1 | 9 | 5 | 3 | 23 | 12 | 5 |
| *Total 1* | *351* | *198* | *44* | *133* | *108* | *22* | *61* | *67* | *10* |
| *Total 2* | *593* | | | *263* | | | *138* | | |

Blue Creek Material by Quality, Social Context and Time Period

Table 3. 55

| Context | Late Formative | | | Early Classic | | | Late Classic | | |
|---|---|---|---|---|---|---|---|---|---|
| | Q-1 | Q-2 | Q-3 | Q-1 | Q-2 | Q-3 | Q-1 | Q-2 | Q-3 |
| E1 | - | - | - | 46.8 | 38.8 | 7.2 | 13.0 | 12.3 | 0.7 |
| E2 | - | - | - | - | - | - | 14.5 | 26.8 | 2.9 |
| NE1 | 4.0 | 8.1 | 3.5 | 0.4 | - | - | - | 0.7 | - |
| NE2 | 1.0 | 1.3 | 0.7 | - | - | - | - | - | - |
| NE3 | 0.2 | - | 0.2 | - | - | - | - | - | - |
| RC1 | 51.8 | 23.3 | 2.9 | - | 0.4 | - | - | - | - |
| RC2 | - | - | - | - | - | - | - | - | - |
| IND | 2.2 | 0.7 | 0.2 | 3.4 | 1.9 | 1.2 | 16.7 | 8.7 | 3.6 |
| *Total 1* | *59.2* | *33.4* | *7.5* | *50.6* | *41.1* | *8.4* | *44.2* | *48.5* | *7.2* |
| *Total 2* | *100.1* | | | *100.0* | | | *99.9* | | |

Percentages of Blue Creek Material by Quality, Social Context and Time Period

Table 3.56

## SECTION 4 -- ALTERNATIVE TECHNIQUES

### INTRODUCTION

As the point of this work is two-fold, to develop a new paradigm for obsidian study and to provide a template of data on which future research may be based, the traditional per-piece analysis has been the focus. This section will briefly address two additional techniques that were investigated for their suitability in furthering obsidian studies. These techniques were analysis of material by weight, rather than per piece, and 'cutting edge to mass ratios'. The material used for this section comes from the Blue Creek collection. A full discussion of these techniques may be found in Chapter 2, Section 1

### OBSIDIAN BY WEIGHT

When the material was discussed simply in terms of source and temporal context (Table 3.1), it was noted that the El Chayal source made up roughly 72 per cent, 83 per cent, and 73 per cent respectively of the entire Late Formative, Early Classic and Late Classic collections analysed by the 'per piece' method. At Blue Creek, using this same 'per piece' method, El Chayal obsidian accounted for 93.7 per cent, 96.8 per cent, and 92.4 per cent respectively of the collections from these same periods (Table 3.19); all are well above the Lowland 'average'.

| | ELC | IXT | SMJ | MX | Total |
|---|---|---|---|---|---|
| LF | 434.8 | 1.6 | 9.1 | 0.0 | 445.5 |
| EC | 69.2 | 3.8 | 2.2 | 0.0 | 75.2 |
| LC | 188.1 | 4.7 | 1.7 | 0.7 | 195.2 |

Total Grams of Each Source by Time Period

Table 3.57

When the Blue Creek material was analysed by weight, the proportions of material altered. In the case of the El Chayal material, the percentages increased for the Late Formative and Late Classic periods (97.6 per cent and 96.4 per cent respectively; Table 3.58) and decreased for the Early Classic Period (92.0 per cent). This suggests that contrary to the 'per piece' analysis, there was actually more El Chayal material available in the Late Formative and Late Classic periods, but that it was spread among fewer pieces.

Material from the other sources also show variations between these types of analyses. San Martin Jilotepeque material, in the previous 'per piece' analysis accounted for 6.1 per cent, 1.2 per cent, and 2.8 per cent respectively of the material from the Formative, Early Classic and Late Classic Periods (Table 3.19). In a 'by weight' analysis this same material composes 9.1 per cent, 2.2 per cent, and 1.7 per cent of these same collections (Table 3.58). When analysed by piece, the data would appear to indicate that material imported from this source increased from the Early Classic to the Late Classic. However, the weight of the material reveals a different pattern, one in which the amount of material from San Martin Jilotepeque declines continuously from the Late Formative through the Late Classic.

| | ELC | IXT | SMJ | MX | Total |
|---|---|---|---|---|---|
| LF | 97.6 | 0.4 | 2.0 | 0.0 | 100.0 |
| EC | 92.0 | 5.0 | 3.0 | 0.0 | 100.0 |
| LC | 96.4 | 2.4 | 0.9 | 0.3 | 100.0 |

Percentage Weight of Each Source by Time Period

Table 3.58

Ixtepeque material when examined per piece shows an increase from the Late Formative through to the Late Classic. The quantities of material increase significantly in each period, starting with 0.5 per cent of the Late Formative collection, 1.9 per cent of the Early Classic material, and 3.7 per cent of the Late Classic material (Table 3.19). However, as with the other material, examination of the material by weight indicates a different pattern (Table 3.58). In the Late Formative, Ixtepeque material accounts for only 0.4 per cent of the collection, a proportion very similar to that by piece. Variations between these two patterns becomes more pronounced in the Early Classic, when Ixtepeque material equal 5 per cent of the collection by weight. Ixtepeque material actually decreases in quantity by weight in the Late Classic, to 2.4 per cent of this collection. This is contrary to the pattern suggested by the 'per piece' analysis, where the Ixtepeque material increases in the Late Classic, and to the traditional opinion of greater reliance on this source in the Late Classic and Terminal Classic periods throughout the lowlands (Dreiss and Brown 1989; Dreiss et al. 1993).

Analysis of material by weight shows a different pattern to the one traditionally predicted for the Maya Lowlands (see Chapter 3, Section 1, this volume). Consequently, future research by weight may be of greater use in determining the quantities of each source being imported into each site, or region, than per piece studies. Such studies however, should not be discounted, as they are excellent indicators of how many people may have had access to material from each source.

## Cutting Edge to Mass Ratios

Sheets and Muto have suggested that a better method by which to analyse obsidian is to determine the usefulness of each piece as a tool (1972). This is accomplished by evaluating the cutting potential of a blade in relation to the amount of the material utilised in its manufacture. Sheets and Muto have put forward the idea that the obsidian may be evaluated in terms of a ratio of the total cutting edges to the mass (CE:M). This technique is useful in determining whether obsidian was 'conserved', that is, if items were manufactured in such a manner as to maximise the usable amount of cutting edge while using the minimum amount of obsidian. A full description of this technique is included in Chapter 2, Section 1.

Sidrys (1979) uses this technique, and its ancillary theory that high CE:M ratios are indicative of material being conserved, to argue for drop off modes of exchange in obsidian trade, noting that average CE:M ratios for lowland sites – those between 160 to 740 kilometres away from obsidian sources – are 5.73:1, while highland sites – those within 15 to 87 kilometres of obsidian sources – are 3.52:1. However, as Crabtree noted, material devoid of inclusions produces finer, thinner blades (1968). This aspect was not addressed in either Sheets and Muto's or Sidrys' discussions.

An examination of material from Blue Creek and Ka'Kabish comparing quality of obsidian – as outlined in Chapter 2, Section 1 – to CE:M ratios revealed a correlation between these two variables (Tables 3.59 and 3.60). As Crabtree's experiments suggested, finer quality obsidian produced blades that were thinner and longer, and therefore with higher CE:M ratios. Declines in these ratios occur concurrently with declines in obsidian quality.

| Quality | Usable Pieces | Avg. CE:M |
|---|---|---|
| 1 | 528 | 153.4 |
| 2 | 346 | 119.9 |
| 3 | 68 | 104.7 |
| Total | 942 | 126.0 |

CE:M Ratios of Different Qualities of Material from Blue Creek

Table 3.59

Consequently, interpretation of data resulting from the application of this technique becomes almost circular; are high CE:M ratios for high quality obsidian the result of this material being physically able to produce finer blades, or, are finer blades being produced from higher quality material in an effort to conserve the material, which therefore results in higher CE:M ratios. Similar circular arguments are applicable to lower grade obsidian, and the decline in CE:M ratios between the qualities of obsidian.

| Quality | Usable Pieces | Avg. CE:M |
|---|---|---|
| 1 | 293 | 195.9 |
| 2 | 297 | 160.3 |
| 3 | 72 | 142.6 |
| Total | 662 | 174.1 |

CE:M Ratios of Different Qualities of Material from Ka'Kabish

Table 3.60

As noted in Chapter 3, Section 3, obsidian sources vary in quality. As El Chayal and Ixtepeque sources are both composed of large proportions of high quality material, it follows that contexts with large amounts of either of these sources would also possess high CE:M ratios. This hypothesis is supported by an examination of the material contextually. Caches, which previously were noted as containing large proportions of high quality material (Table 3.54), also possessed high CE:M ratios (Table 3.61). Conversely, contexts which possess larger proportions of middle to low grade obsidian, such as the Late Formative NE1 deposits (Table 3.62), have lower CE:M ratios.

However, as noted above, it is virtually impossible to determine the intent behind the production of fine blades, and therefore, whether the CE:M ratios reflect the natural fracturing ability of the obsidian, or an effort to conserve the material. Consequently, although this technique may prove to be useful in the future, this work, as noted above, focuses on presenting a contextual model for obsidian research that may provide groundwork for future studies. As such, and in view of the problems inherent in CE:M ratio analyses, this technique was viewed as having limited current application.

| A-Context | LF | EC | LC |
|---|---|---|---|
| C | 167.2 | 190.5 | 69.0 |
| CF | 87.8 | 89.1 | 54.6 |
| CF/COL | | | |
| COL | | 18.9 | 55.2 |
| FF | 33.4 | 103.8 | 74.6 |
| FS | 121.5 | 64.4 | 54.7 |
| H | | | 66.0 |
| H/CF | | | 73.2 |
| H/COL | | | 91.3 |
| H/FS | | | 23.9 |
| M | 80.9 | | |
| MG | 112.7 | 51.9 | |
| OTH | 63.8 | 74.3 | 36.8 |
| PZ/S | 44.0 | | 98.0 |
| SD | | | 54.3 |
| | 150.7 | 180.5 | 58.3 |

Average CE:M Ratios by Archaeological Context and Time Period

Table 3.61

| S-Context | LF | EC | LC |
|---|---|---|---|
| E1 | | 190.4 | 44.3 |
| E2 | | | 62.0 |
| NE1 | 80.5 | 54.7 | 98.0 |
| NE2 | 92.2 | | |
| NE3 | 178.9 | | |
| RC1 | 167.8 | 71.5 | |
| RC2 | | | |
| IND | 68.2 | 78.2 | 64.8 |
| | 150.7 | 180.5 | 58.3 |

Average CE:M Ratios by Social Context and Time Period

Table 3.62

## Chapter 4
## Patterns of Obsidian Consumption and Distribution

### Introduction

It has been established, both in this volume and in numerous site reports, that obsidian source utilisation changes through time. Material from the San Martin Jilotepeque source in Guatemala appears in the earliest Middle Formative Period deposits. El Chayal obsidian from the Guatemalan Highlands is introduced in the Late Formative, where it comes to dominate the collections. Obsidian from other sources, including Ixtepeque, are also present in this period. In the succeeding Early Classic Period, obsidian from the El Chayal source continues to dominate most collections, although a greater diversity of obsidian, including material from sources in Central Mexico, was available. In the Late Classic Period, although a wide variety of obsidian was available, El Chayal and Ixtepeque obsidian dominate the collections. It is during this period that a second shift in source utilisation occurs, away from El Chayal in favour of Ixtepeque.

The pattern of shifting source utilisation has been well documented. Reporting on obsidian using this source and time model has become the standard in obsidian research in theMaya area. While the establishment of this pattern has been instrumental in forming the foundation for what we understand about obsidian source utilisation, it is my opinion that it is time to take the next step and look for reasons behind this pattern.

The research presented in this work was formulated around a series of very simple questions. As it is apparent that there was a transition from San Martin Jilotepeque material to El Chayal in the Late Formative, and another from El Chayal to Ixtepeque in the Late Classic, the natural question is why some sources appear more frequently or exclusively in some contexts. If a pattern exists, does it change through time, and, if so, why? Can such a pattern account for the transition between sources? These questions are designed to gain a better understanding of who was using which types of obsidian, and for what purpose were these different obsidian types were used. It is my opinion that the answers to these questions are fundamental in understanding why different obsidian sources were used. The following sections will summarise the data presented in the previous chapter, discuss the relevance of the patterns discovered, and propose avenues of future research.

## Section 1
## Summary of Data

### Temporal Summary

The transition from San Martin Jilotepeque to El Chayal material has been documented at many sites in the Maya Lowlands (Dreiss 1989; Dreiss and Brown 1989; Dreiss *et al.* 1993; Moholy-Nagy and Nelson 1987, 1990). This pattern was confirmed by the data analysed in this volume (Chapter 3, Section 1, Table 3.1). While material in this volume show a dramatic change in the Formative Period where San Martin Jilotepeque material drops from 71.6 per cent to 11.9 per cent of the collection from the Middle to Late Formative, it was discovered that this decline is deceptive in terms of trade in San Martin Jilotepeque material. Although the proportion of the material did decrease, this was discovered not be a factor of declining importation of material, but rather an increase in the importation of the El Chayal material. It was discovered that while the proportion of San Martin Jilotepeque material declined, reflecting the introduction of other obsidian sources, the actual amount of San Martin Jilotepeque material increased, rising from 48 pieces reported in the Middle Formative to 67 pieces in the Late Formative.

The shift was most noticeable among the river sites, where San Martin Jilotepeque was the only material reported in the Middle Formative Period. In this area, the proportion of San Martin Jilotepeque declines from 100 per cent in the Middle Formative Period to six per cent in the Late Formative. El Chayal however, goes from not being represented in the Middle Formative to comprising 80.6 per cent of the Late Formative collection. However, as with the general collection, these percentages mask an increase in the amount of San Martin Jilotepeque obsidian being imported in the river region.

The marked increases in Ixtepeque material in the Late and Terminal Classic Periods are also noteworthy. A second shift in obsidian consumption, away from El Chayal material in favour of Ixtepeque obsidian in the Late Classic, has also been noted in the literature (Dreiss and Brown 1989; Dreiss *et al.* 1993). Again, this pattern was confirmed by research in this volume (Chapter 1, Section 1, Table 3.1).

### Functional Analysis Summary

The first type of contextual analysis conducted on the material collected for this study was by functional contexts. This study was a broad based analysis designed to locate fundamental patterns that could be investigated in greater detail in the more comprehensive archaeological and social context analyses. A definite pattern quickly became apparent in the functional analysis; El Chayal obsidian was the primary, if not exclusive, material used in rituals, while San Martin Jilotepeque material was restricted to waste or domestic deposits.

The only material reported from the Middle Formative Period originated in waste deposits. The lack of contextual variety makes it unwise to derive context-based patterns for this period.

In the Late Formative Period, obsidian was recovered in all five contexts: Ritual, Waste, Domestic, Burial, and Indeterminate. The majority of obsidian (58.0 per cent) was concentrated in Ritual deposits. Of note is that all of this material is El Chayal in origin. Both San Martin Jilotepeque and Ixtepeque obsidian are also present in this period. The majority of San Martin Jilotepeque (93.9 per cent), and Ixtepeque (94.6 per cent) material is concentrated in Waste deposits. The remaining pieces are from either Domestic or Indeterminate contexts (Table 3.6).

Ritual deposits in the Early Classic Period are still dominated by El Chayal obsidian, although it is no longer the exclusive source used for these deposits. El Chayal accounts for 97.4 per cent (n = 189) of the material recovered in ritual contexts. The other 2.6 per cent (n = 5) is spread between Mexican obsidian (1.5 per cent, n = 3), and Ixtepeque and San Martin Jilotepeque obsidian (0.5 per cent each, n = 1 each). The single piece of San Martin Jilotepeque material is the only piece from this source recovered in a ritual context in all five time periods. The remaining pieces of San Martin Jilotepeque obsidian were all recovered from Waste contexts. The Ixtepeque and Mexican obsidians were deposited in Waste, Domestic, and Burial contexts, although the majority of pieces from both sources (71.4 per cent and 69.6 per cent respectively) were recovered from Waste Contexts.

Although El Chayal obsidian continued to dominate the Ritual deposits of the Late Classic Period, accounting for 98.7 per cent of the material found in these contexts, this period saw a decline in ritual consumption of obsidian. During the preceding Late Formative and Early Classic Periods, ritual activity consumed 58.0 per cent and 60.7 per cent of the obsidian collections respectively for these periods. In the Late Classic Period, this amount decreases to 36.2 per cent of the total collection, 35.8 per cent being El Chayal material and 0.2 per cent each coming from Ixtepeque and Mexican sources. As in the earlier periods, the majority of San Martin Jilotepeque material (90.0 per cent) is concentrated in Waste deposits, with none found in Ritual contexts.

The decline in ritual consumption of obsidian is complete in the Terminal Classic Period, with no such contexts being reported. The most noticeable trend in this period is the increase in Ixtepeque material, which accounts for 41.9 per cent of the collection. This material dominates the Waste contexts. Both Ixtepeque and El Chayal material are reported in Burial contexts, and it is here that the latter source still dominates.

In summary, the pattern that emerges is one where El Chayal material appears to be the material of preference for Ritual and Burial deposits, in all periods. That it is concentrated in these contexts suggests that an intra-site mechanism, social, political, or ideological in nature, may have been responsible for diverting this material out of general circulation. However, lesser amounts of this material are found in other contexts during the Late Formative and Early and Late Classic Periods, suggesting that a 'trickle down' phenomenon was in effect.

That San Martin Jilotepeque material is, with one exception, not found in either Ritual or Burial contexts is also of significance. It suggests that not all obsidian was viewed equally, and that for some reason this material may have been deemed ritually, or ideologically, unacceptable or inappropriate.

### Archaeological Analysis Summary

Once the general patterns of Ritual preference of El Chayal material, and restriction to Waste contexts of San Martin Jilotepeque obsidian, were determined, an investigation of the archaeological contexts was conducted to illuminate the nuances of these patterns. Obsidian from the earliest period, the Middle Formative, was all recovered in construction fill contexts. This restricts the inferences that can be made regarding patterns of disposition. It is probable that at least some of this material originated in other contexts and was recycled. The incorporation of midden deposits and other domestic refuse into construction fill is relatively common in the Maya area. The scarcity of deposits from this period also make it difficult to ascertain with any confidence a pattern of obsidian consumption for this period.

A wider variety of contexts were identified from this period. The predominant ritual activity was caches, in which El Chayal obsidian was the exclusive material; it was also the exclusive material recovered from mortuary contexts (Table 3.20). Other sources, San Martin Jilotepeque and Ixtepeque, cluster in construction fill (n = 28 and 31 respectively), followed by midden contexts (n = 33 and 1 respectively), both of which were classed as 'waste'.

In the Early Classic period, ritual activity is still concentrated on caching, although El Chayal is no longer the only material included in such deposits. Mexican obsidian appears for the first time during this period, and small amounts of it are present in both caches and burials (n = 3 each). The largest concentration of this material (n = 8; 34.8 per cent) was found in construction fill contexts.

During the Late Classic Period, ritual activity shifts away from caches to 'special deposits'. Caches, although still present, use only 3.7 per cent of the obsidian deposited, while special deposits accounted for 11.3 per cent of the collection. Although more popular than caches during this period, special deposits still represented a much smaller proportion of the Late Classic Period obsidian collection than caches did in the earlier periods (Table 3.21). In preceding Early Classic and Late Formative Periods, caches utilised 60.7 per cent and 58.0 per cent of their respective collections. Workshop deposits were also identified during this period, and the material contained therein was split roughly evenly between El Chayal and Ixtepeque obsidian (52.9 per cent and 47.1 per cent respectively). Deposits in the Terminal Classic were limited to four accurately identifiable contexts, construction fill, mortuary goods, middens, and workshops. The first two contexts were dominated by El Chayal obsidian, while Ixtepeque material dominated the latter one.

### Social Analysis Summary

During the Late Formative Period, the largest concentration of obsidian was identified in ritual/ceremonial contexts. These contexts contained 57.8 per cent of the obsidian recovered in this period. The highest social group identified was Non-Elite 1. This group consumed 12.1 per cent of the obsidian collection. Material utilised by this group was roughly evenly split between El Chayal and San Martin Jilotepeque obsidian (52.9 per cent and 47.1 per cent respectively). Proportions of El Chayal and San Martin Jilotepeque obsidian in the NE2 group were weighted towards consumption of El Chayal obsidian (77.8 per cent). However, this may the result of a small sample (n = 9). McKillop noted that in small samples such as this, there is a noticeable bias towards El Chayal material (McKillop 1996).

In the Early Classic period, the highest status group was identified as Elite 1. This group used 43.9 per cent of the obsidian reported from this period. This was the largest concentration of obsidian from this period. It is also denotes an increase of roughly 350 per cent over the consumption of obsidian by the NE1 group in the Late Formative. This increase was not restricted to the highest status section of society, and in the Early Classic period a corresponding increase of 360 per cent in obsidian consumption by the non-elite groups was also noted. These increases were not matched by public ritual/ceremonial contexts, where consumption decreased during this period.

In the Late Classic period, the consumption of obsidian by non-elites is almost triple that of the previous Early Classic period, accounting for 14.6 per cent of the Late Classic collection where it had previously been only 5.3 per cent of the Early Classic collection. Consumption of obsidian among the elites decreased slightly from 44.1 per cent of the material in the Early Classic to 33.4 per cent in the Late Classic period. San Martin Jilotepeque obsidian was found in limited quantities in both elite and non-elite contexts, where it accounted for 1.5 per cent and 5.2 per cent of the respective collections. No San Martin Jilotepeque material was recovered from either ritual or workshop contexts.

Workshop debris increased in the Terminal Classic, where it accounted for 15.2 per cent of the collection, up from 4.3 per cent in the previous period. Obsidian debris recovered from this context was El Chayal and Ixtepeque in origin. The limited amount of San Martin Jilotepeque that was recovered from this period (3.4 per cent) was divided evenly between Elite 1 and Indeterminant contexts.

The primary pattern noted for the consumption of obsidian in this analysis was the gradual decline in public ritual in favour of elite, or private, consumption. In the Late Formative period, the greatest consumption of obsidian was in ritual contexts of a public nature. In the Early Classic through the Terminal Classic, this pattern shifts, with elites forming the largest consumer group. Public ritual activity declines steadily through these periods. Non-elite consumption is highest in the Late Classic period.

### Contextual Analyses Summary

El Chayal obsidian was the primary, and in some cases exclusive, obsidian used in ritual activities. This pattern holds true for both public ritual deposits as well as those recovered from more private locales, such as domestic structures or residential courtyards. A single piece of San Martin Jilotepeque obsidian was recovered in a ritual context, a cache located in a private residential courtyard adjacent to the core area of the Blue Creek Ruin.

Distribution of San Martin Jilotepeque material was limited to waste or domestic contexts. The single example noted above forms the exception to this rule. Mexican obsidian, despite its more remote origin, did not appear to be more highly coveted in any manner. It was not reserved for ritual or elite purposes, and the majority of the material was recovered from waste contexts.

Ritual activity appears to peak in the Early Classic period, where it accounted for 60 per cent of the material, up slightly from the Late Formative, where it accounted for 58 per cent of the collection. In the Late Classic, ritual activity not only declines, utilising only 36.2 per cent of the obsidian collection, but also shifts in orientation, away from single cache events, towards dispersed special deposits. The limited evidence from the Terminal Classic suggests that this pattern of decline may continue, however, the size of the collection warrants caution in assuming patterns. What is of note is that ritual activities which use obsidian, whether cache or special deposits, appear to shift away from public venues in the Early Classic, towards deposits interred in private, or residential, locations.

## SECTION 2
## DISCUSSION

One of the questions asked at the start of this chapter was "do some sources appear more frequently, or exclusively, in some contexts". It is clear from the data presented in this work that the answer to this question is yes. Obsidian from El Chayal and San Martin Jilotepeque demonstrate context related patterns. The former source is the predominant, if not exclusive, source used in caches and rituals, while the latter source is restricted to domestic and waste deposits. The other sources, Ixtepeque and Mexican, do not seem to be affected by the 'glass ceiling' that restricts the San Martin Jilotepeque material, although neither of these two sources come close to the quantities of El Chayal material in ritual contexts. In fact, the majority of both of these sources were located in waste or domestic deposits.

The next question is 'why do these materials appear in these contexts'. The material was investigated to determine if preference for El Chayal, and conversely, prejudice against San Martin Jilotepeque, for ritual purposes might be tied to the quality of the material. Quality here is assessed on its ability to produce smooth, elegant blades, an ability which Crabtree notes is directly linked to the size and quantity of inclusion in the material (Crabtree 1968). An examination of the material from the Blue Creek Ruin and Ka'Kabish, the only sites available for such study, indicated that it was possible to distinguish different 'grades' of obsidian based on inclusions (see Chapter 2, Section 1, and Chapter 3, Section 4, this volume). It was also determined that material of all grades was available from each obsidian source. The bulk of the El Chayal material was deemed to be of high quality (68.3 per cent), while middle quality obsidian comprised the largest type of San Martin Jilotepeque material (50 per cent), and low quality material the next type (38.9 per cent). However, when these characteristics were compared with contextual data, specifically in the case of ritual deposits, it was discovered that quality was not affected by the contexts. Material of all qualities was recovered from ritual contexts, as was the case for domestic and waste contexts. Since the El Chayal material had a higher proportion of high quality material, the high proportion of high quality material in deposits where El Chayal dominated was expected. However, both middle and low qualities of this material were located in these deposits as well. Closer examination of the material by source, quality, and context (Table 3.54) could determine no discernible pattern in preference for other qualities of obsidian in contexts. Consequently, it may be assumed that quality, as defined here, was not a contributing factor in obsidian source distribution. A point of note however, is that the one piece of San Martin Jilotepeque obsidian included in the Elite 1 cache was of low quality material.

Consequently the questions still remains: why does El Chayal obsidian come to dominate obsidian collections; why was it the preferential, and occasionally exclusive, material for ritual deposits; and why, when other material does appear in ritual contexts, albeit in limit quantities, does San Martin Jilotepeque obsidian seem to be proscribed from such contexts?

The answer often given for the first question lies not in the policies of intra-site consumption but in the politics of the Maya Highlands. The site of Kaminaljuyu, approximately 25 kilometres to the south of the El Chayal obsidian source, becomes a major centre in the region at the end of the Late Formative and remains so throughout the Early Classic (Brown 1977). The rise in prominence of El Chayal obsidian has been correlated with the rise in ascendancy of Kaminaljuyu. It has been postulated that this site, whether independently or at the incitation of Teotihuacan, attempted to control the obsidian trade networks (Sanders 1973:352-353; Sidrys and Kimberlin 1979). Evidence from Kaminaljuyu indicates that obsidian mining and artefact production were major activities at the site and large deposits of workshop debitage have been reported (Sanders 1973:352). The predominant obsidian source identified at Kaminaljuyu was El Chayal. This source accounted for 95 per cent of the material identified (Sidrys and Kimberlin 1979).

Kaminaljuyu has also been viewed as participating in a trade triumvirate with Teotihuacan and Tikal (Arnold 1990). Tikal has long been viewed as one of the pre-eminent sites in the Maya realm (Coe 1993; Harrison 1999; Proskouriakoff 1993; Rathje 1977; Schele and Freidel 1990; Sharer 1994). If one accepts the premise that in the Early Classic, the site of Tikal was tied not only economically through exchange networks with Teotihuacan, but also by blood (either through marriage or occupation), and that Kaminaljuyu was also tied to Teotihuacan through similar means, then the choice of obsidian among these sites may have been part of this solidarity. If one furthers this hypothesis, Tikal, being a well respected site, may have either served as a redistribution node for the El Chayal obsidian, or merely had its consumption patterns emulated by sites wishing to partake of this status connection. Furthermore, the decline in El Chayal obsidian at the end of the Early Classic period may therefore be viewed as a side effect, or consequence, of the corresponding decline in authority of Tikal at this point in time. Hence the choice of El Chayal material may indicate subtleties of status and ideology hitherto unexplored.

Returning to the issue of Kaminaljuyu 'controlling the trade networks', it is my opinion that a difference may be perceived between 'control' of trade networks and 'domination' of these networks. The first implies that material only traversed these route under the auspices of, or with the permission of, a particular site, in this case Kaminaljuyu. The latter implies that a site was able, through mass production, or high demand for a product they produced, to flood the trade routes. This latter situation seems more plausible, as the San Martin Jilotepeque obsidian source continued to be worked, and the material traded, throughout the Classic period. As this study has demonstrated that while the proportions of this material may have decreased in the Late Formative and Early Classic, this was due to an increase in the production and distribution of El Chayal, and that the actual quantities of San Martin Jilotepeque increased in the Late Formative Period, where they peaked (Table 3.1). This material continued to contribute to the obsidian exchange system throughout the Classic period in fluctuating quantities. This supports Braswell's study, which found the San Martin Jilotepeque area was occupied from the Middle Formative through to the Post-Classic (Braswell 1996:696-706). Braswell also notes that the obsidian production from this region never exhibited a high level of political control, or macro-regional integration, and that production and exchange of this material was probably at a household level. Distribution was most likely through a process of 'down-the-line' exchange networks, rather than a formal system of trading (Braswell 1996). Consequently, the influx of El Chayal obsidian through the trade networks becomes less a matter of 'control' than an ability, through managed production and organised exchanged, to increase supply and accommodate demand for obsidian in the Lowlands.

This raises the issue of the nature of economic organisations at an inter-site and intra-site level. The two competing theories of economic organisation in the Maya realm centre on concepts of market system and political economies (Lucero 1999; McKillop 1996; Mitchum 1994). Polanyi describes a market economy as a "a self-regulating system of markets . . . an economy directed by market prices and nothing but market prices" (Polanyi 1957:43). The term 'political economy' has never been adequately or explicitly define in Maya literature. According to Earle, political economies may be interpreted as "the calculated manipulation of competing chiefly factions" (Earle 1997:45). In Maya literature, especially those articles dealing with Classic period economies, the concept of political economy implied focuses more on the "calculated manipulation" or control aspects, and less on the factional 'competition' of chiefs (Freidel 1981b; Ford 1991). It is with this definition in mind that the term political economy will be used.

The value of economic resources in the development of social hierarchies has long been accepted (Demarest 1992; Earle 1997; Fried 1960; Hayden 1995; Lucero 1999). Earle considers control of resources, along with military power and ideology, the primary means by which social stratification may occur (Earle 1997). These resources may have been dispersed to create a system of status based on obligation (Freidel 1979; Hayden 1995; Mauss 1994), retained to form status symbols (Bohannan 1966; Rappaport 1968), or used in elaborate ritual displays of authority (Freidel 1979; Guderjan and Weiss 1995; Lucero 1999). In all of these scenarios, the power of material objects to create or affirm authority is evident.

Brumfiel notes that "valuables acquired from distant sources supply considerable political control because of their ability to attract followers, allies and patrons and to maintain hierarchies of control" (Brumfiel 1994:6). This idea implies that an item is valued because of the difficulty of procurement, and is coveted due to its uniqueness. This concept would therefore apply on a broad scale to all the materials imported over an equally great distance. However, if such were the case, then both San Martin Jilotepeque and El Chayal material would be equally valued. To extend this rationale, the value of a good may be seen as directly related to the distance or difficulty which was surmounted in its procurement. Hence, obsidian from Central Mexico, which was traded over greater distances than material from the Maya Highlands, would be of proportionately greater value. If value translates as the manner in which a material was used, then we would expect that Mexican obsidian would appear predominantly in ritual or elite contexts. Research presented in this volume demonstrates that this is clearly not the case, as larger quantities of this material were recovered from waste deposits than from ritual.

Consequently, it remains that material from the El Chayal source was being chosen deliberately for reasons other than distance or value. One may argue that the lack of organisation involved in the production and distribution of San Martin Jilotepeque material would mean that obsidian from El Chayal would have been easier to obtain and therefore was more likely to appear in ritual or other deposits where a demand for large quantities of obsidian was needed. If such was the case, then why was El

Chayal not used exclusively in all ritual, or high demand, deposits? In the Early Classic period through the Terminal Classic period, Ixtepeque and Mexican obsidian were used in caches, special deposits, and burials, albeit in the first two instances the quantities are very limited. San Martin Jilotepeque material appears only once in an Early Classic cache located in a private residential courtyard complex.

Walker theorises that material contained in cache deposits may not be simply offertory items, but objects used in rituals, and therefore imbued with a certain sanctity, for which disposal in a common waste deposit would be unthinkable (Walker 1995). It is possible that large deposits of obsidian may be 'ceremonial trash' from blood letting rituals. Freidel and Schele (1988) liken obsidian blades to "ams", small stinging black spiders with red spots (blood drops), viewing obsidian as an important ideological component of kingship rituals. Large cached deposits of blades may be the by-products, or ceremonial trash, resulting from bloodletting. Hence, the predominance of a single type of obsidian in these deposits may be representative of a more controlled type of ritual. However, the variations in the Barium rates of the El Chayal obsidian contained in these deposits suggests that the material came from more than one location, or quarry, in the El Chayal source. If the material used for bloodletting rituals was viewed as metaphors for these 'black spiders', then is it possible that colour may have played a role in the selection of material, explaining the preference for El Chayal, which is predominately grey to black, and not for Ixtepeque material, which is honey-brown. Pachuca obsidian, which is green, may also fall into this exclusionary range, although this is strictly conjecture, as this study did not include Pachuca material as a separate category, due to the limited size of the sample. Furthermore, it should be stated that while this 'colour selection' concept is strictly conjecture, it is based on material that would have been used for bloodletting and not for other, perhaps strictly offertory, obsidian material.
Attempts to identify those caches of obsidian which contain the residue of ritual activities versus those which contain offertory material may lead to an investigation of any additional material contained within the caches. Of the 66 obsidian cores discovered at the Blue Creek Ruin, 62 (93.9%) were recovered from large caches of obsidian blades. All were in various stages of being expended, some were bi-directional in nature, and one fragment was the discarded proximal tip from a rejuvenated core (*cf.* Andrefsky 1998:11-17). Of the remaining pieces, three were broken, utilised core sections recovered from construction fill (n=2), or humus (n=1). The remaining piece was a small, distal tip discovered from a burial.

Studies by Hay (1977) and Lewenstien (1981) have shown that some obsidian was used for utilitarian purposes, supporting the belief that obsidian was used for both functional and utilitarian purposes. One may assume that at least some of the material in midden or domestic contexts may have been used for utilitarian purposes. It is possible that some material in domestic contexts was not used for everyday utilitarian purposes, but retained for specific, and perhaps special events. Lewenstein's study of use-wear of Cerros obsidian does not differentiate between types of obsidian, merely noting that the artefacts were all 'grey' (Lewenstein 1987:140). It would be interesting to know whether the different sources were used for different purposes, or if one source was used more often for utilitarian purposes.

Consequently, on returns to the questions of why El Chayal obsidian dominates in ritual and mortuary contexts, and why San Martin Jilotepeque material is found predominately in waste contexts. However, there are now avenues to explore: examinations of selection based on trade networks (reliability, exchange organisation, and trade partners), and association of material ideologically.

## SECTION 3
## CONCLUSIONS

### CONCLUSIONS

This report has demonstrated that a pattern may exist between obsidian source utilisation and consumer contexts. While the data presented in this report demonstrates a strong correlation between El Chayal and ritual contexts, and San Martin Jilotepeque material and waste contexts, the previous statement regarding the presence of these patterns is prefaced with a 'may'. The reason for this is simply that obsidian reports rarely include the contextual data of the pieces discussed, limiting any contextual analyses due to database size.

The purpose of this work was not to provide absolute answers regarding obsidian consumption and distribution patterns, but rather to present a new analysis model that may further our understanding of such patterns. With the advent of more accurate sourcing methods, coupled with refinements in dating techniques, we are now able to securely identify both the source and temporal period for obsidian artefacts. Indeed, research along this vein has dominated the obsidian literature. I would suggest that this type of source/time analysis is no longer sufficient; it is time to take obsidian analysis to the next level, to examine the identified patterns of obsidian source consumption with an eye toward the use of material from these sources at an intra-site level.

### FUTURE ISSUES

In furtherance of the idea of a new analysis model, it is incumbent to discuss issues and possible avenues of future research that, while beyond the scope of this work, may comprise the next level of obsidian analysis. The first issue that must be addressed is the matter of obsidian reporting. A complete and rigorous system of reporting the total quantities of obsidian recovered and their contexts is warranted if the patterns outlined above are to be fully investigated. With a more complete picture as to the distribution of obsidian throughout a site, better research designs may be formulated. Testing of artefacts may be streamlined to reflect specific deposits, or, ideally, use sampling strategies that provide an equal, proportional, representative sample from each obsidian deposit. This latter technique would further investigations into intra-site distibution patterns and eithervalidate, or refute the pattern of obsidian consumption and distribution noted in this work.

Future studies may also wish to investigate the variations noted in this study between the number of pieces present at a site versus the total amount present. The former is useful in determining how many people may have had access to the material, but the latter is a better indicator of the volume of material being imported. As noted at the Blue Creek Ruin, these amounts are often different. Between the Late Formative and Early Classic periods the number of piece of El Chayal material in the samples increased from 93.7 per cent 96.8 per cent. While this may be interpreted as an additional 3.1 per cent of El Chayal material being available, the actual volume of the material available declined from 97.2 per cent to 92.0 per cent in the same time. Likewise, the decline in number of pieces to 92.4 per cent in the Late Classic was not matched by a decline in volume, but an increase to 96.4 per cent. Consequently, while per piece analyses are valuable indicators of how many people may have had access to the material, inclusion of the weight of the material would provide better information regarding the volume of material being traded.A further avenue of exploration may be investigations into the specifics of which sites, or which areas, had the highest quantities of El Chayal material in caches. If, as speculated above, El Chayal material was selected for rituals due to its association with Tikal, Kaminaljuyu, and/or Teotihuacan, then do sites that are allies of this triumvirate demonstrate larger proportions of, or exclusive use of, El Chayal obsidian in their rituals in the Early Classic? Unfortunately, material from Calakmul, Caracol, or the other sites speculated as having participated in the defeat of Tikal at the end of the Early Classic was not available for analysis. However, when this material becomes available in the future, such an investigation would prove interesting.

## Last Words

Research in archaeology is subject to rapid changes, often between one season's discoveries and the next. Models, or theories, that seem valid may change over the course of five or ten years. However, using the data currently available I believe that I have successfully demonstrated that enough of a correlation exists between obsidian sources and contexts, functional, archaeological, and social, to warrant researchers taking the next step and considering context as a valuable factor in studies of obsidian consumption and distribution patterns.

**APPENDIX I**

**MAPS, FIGURES, AND PHOTOGRAPHS**

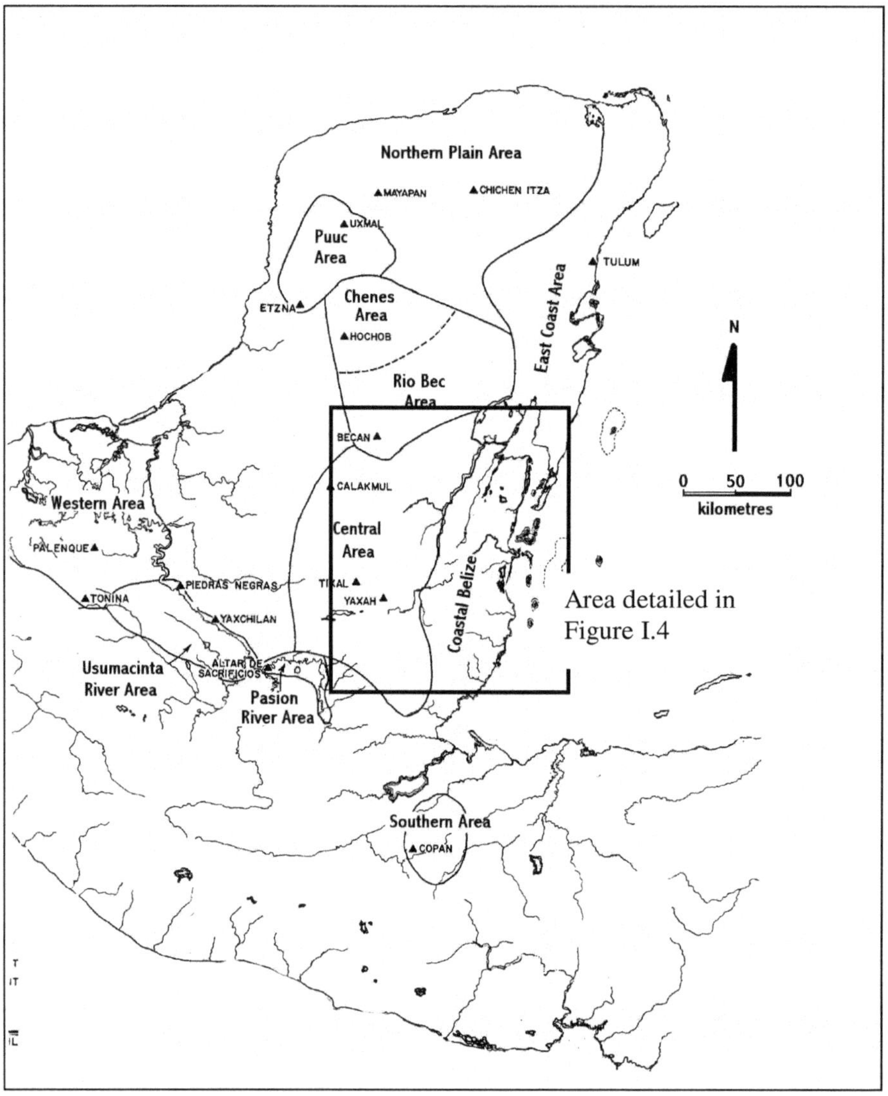


Map of Yucatan Showing Sub-Divisions

Figure I.1

*(adapted from Pollock 1965: figure 1)*

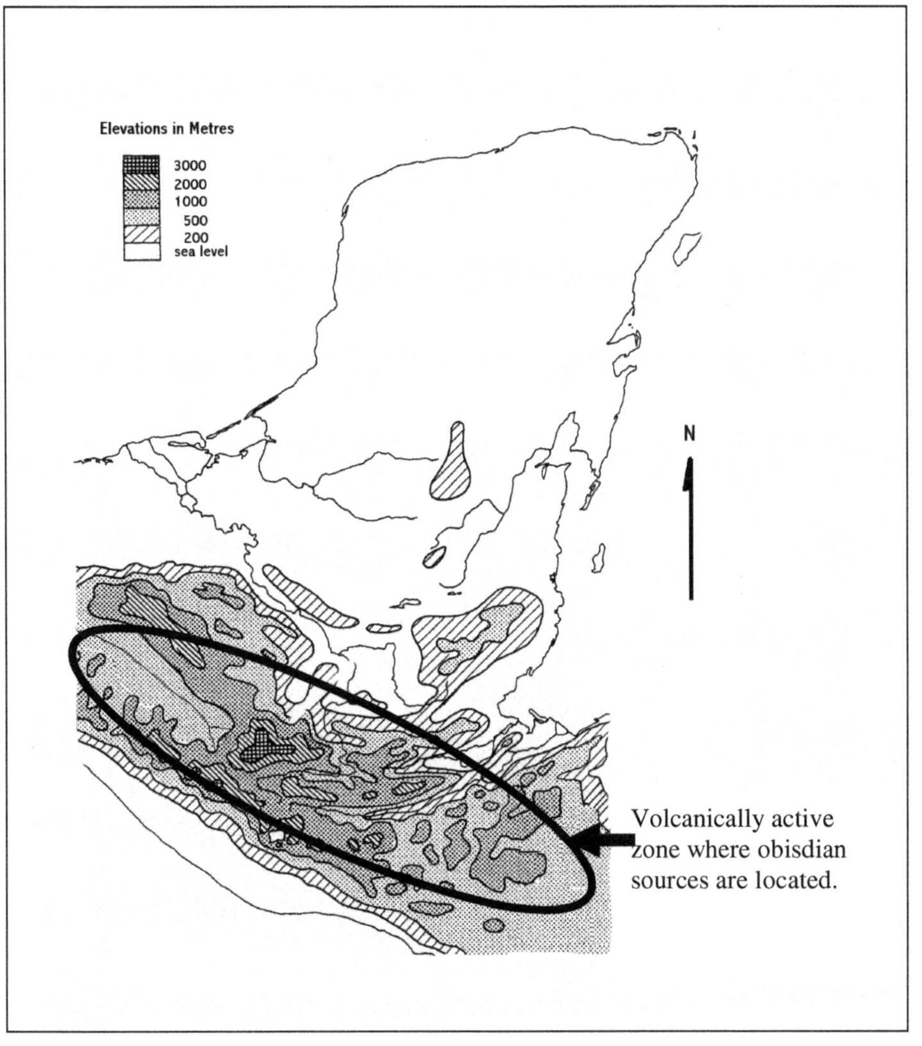


Map of Yucatan Showing Topography

Figure I.2

*(adapted from Sharer 1994: figure 1.2)*

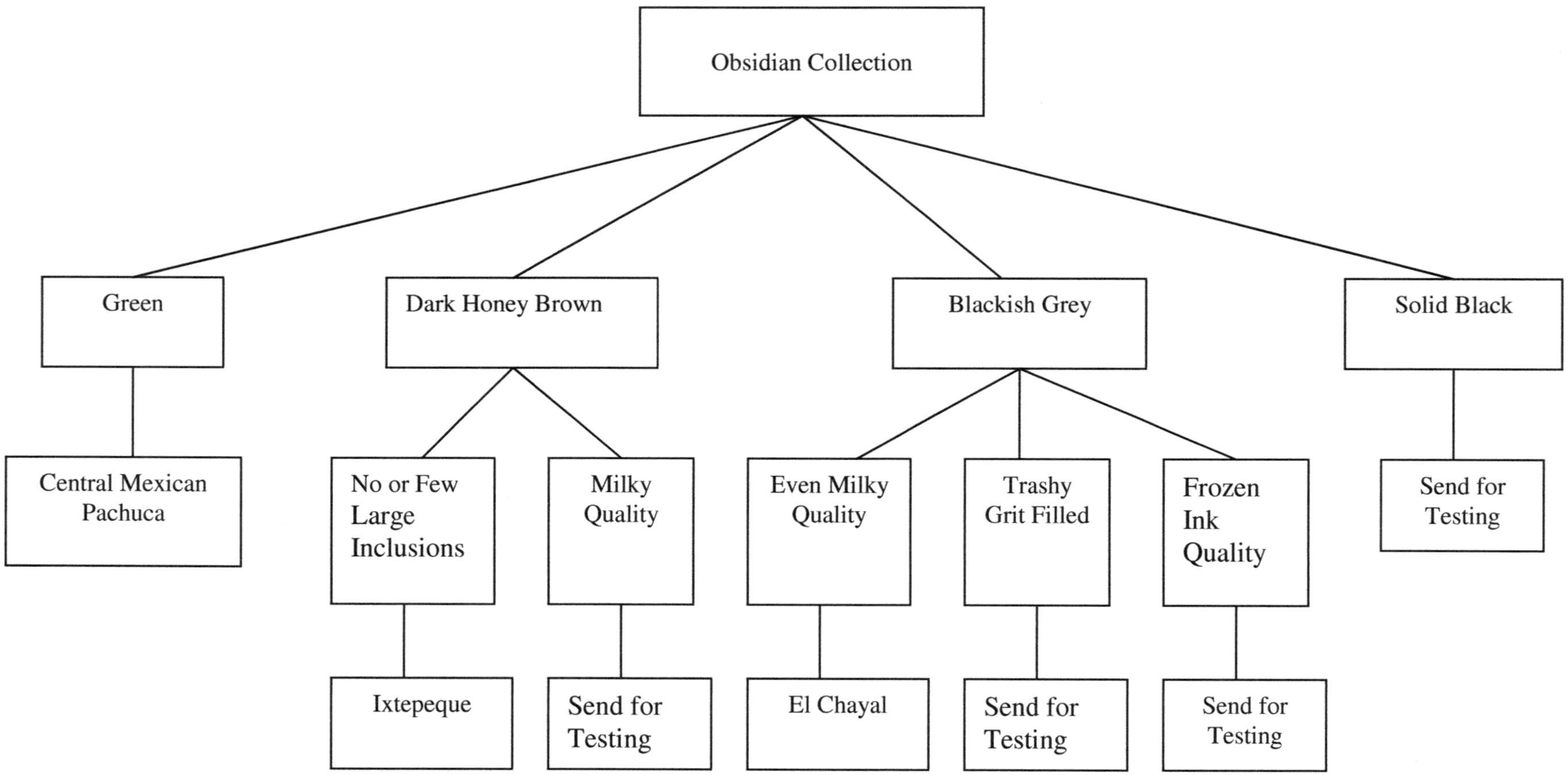


Process for Visually Culling Artefacts

Figure I.3

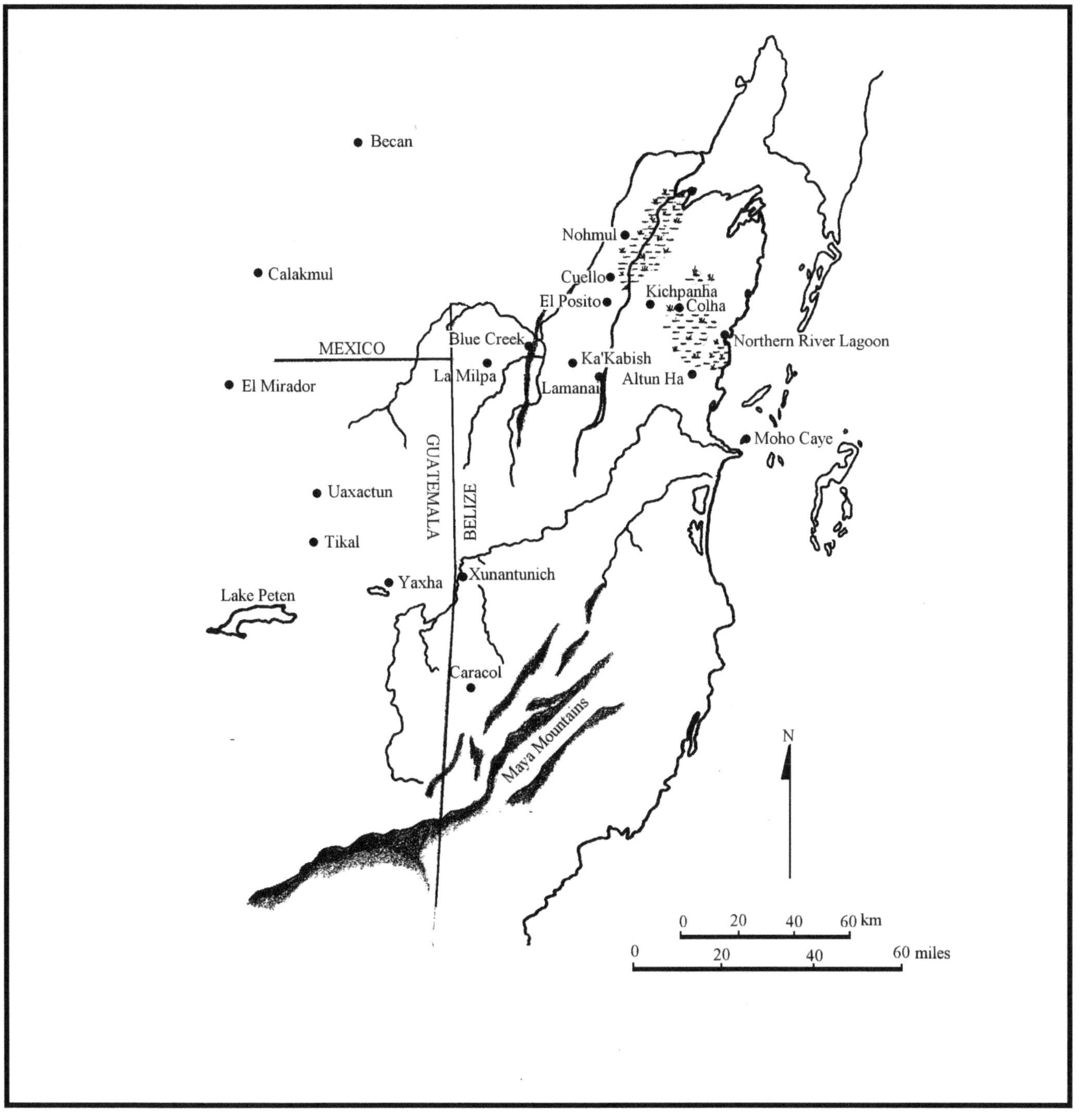


Map Showing Location of Sites in Study Area

Figure I.4

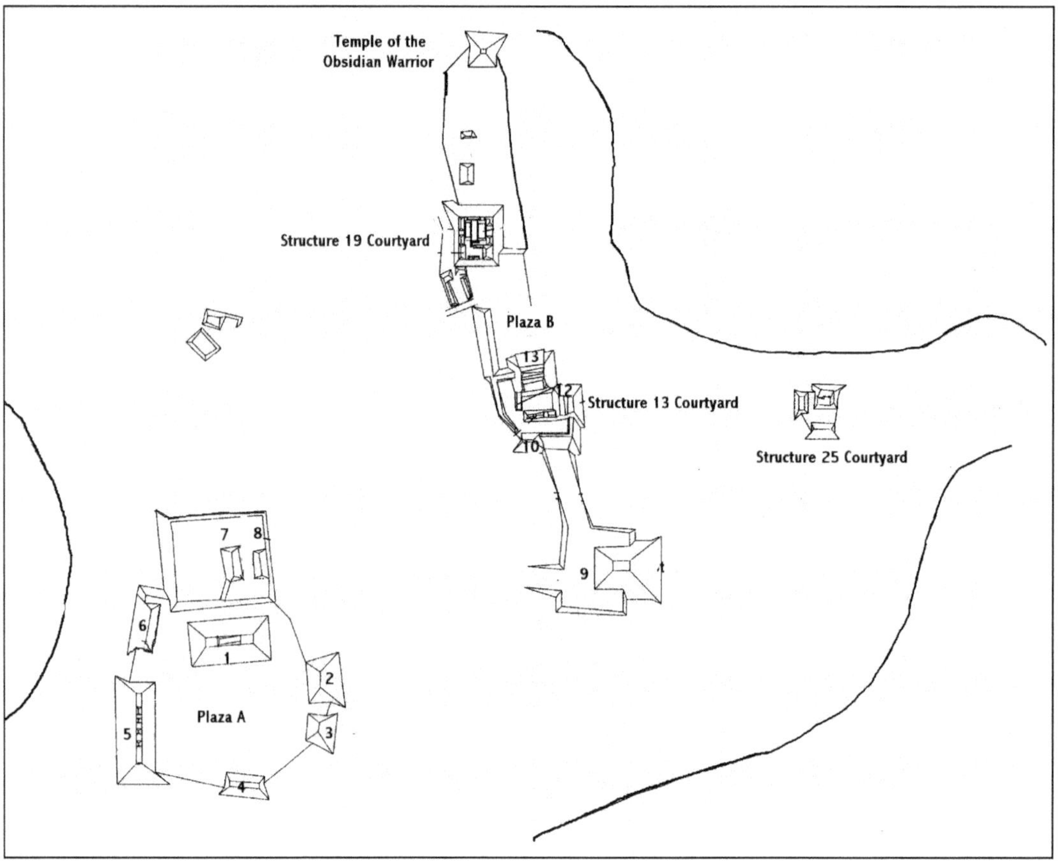


Site Map of Blue Creek Ruin, Orange Walk District, Belize

Figure I.5

*(adapted from Guderjan and Driver 1995)*

Photograph of Obsidian Blades

Figure I.6

Photograph of Obsdian Biface

Figure I.7

*(distal end)*

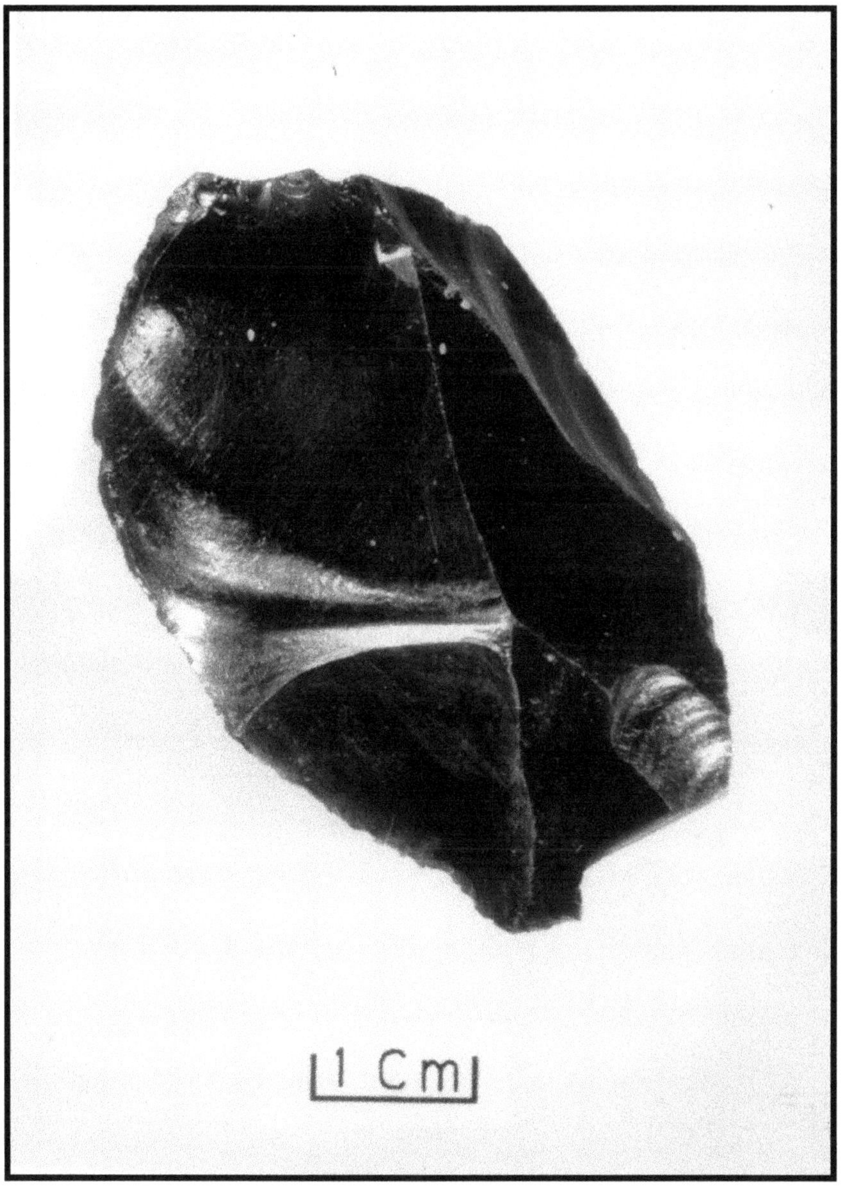


Photograph of Obsidian Uniface

Figure I.8

Photograph of Obsidian Decoration

Figure I.9

Photograph of "Frozen Ink" in Obsidian

Figure I.10

Photograph of "Wisps" in Obsidian

Figure I.11

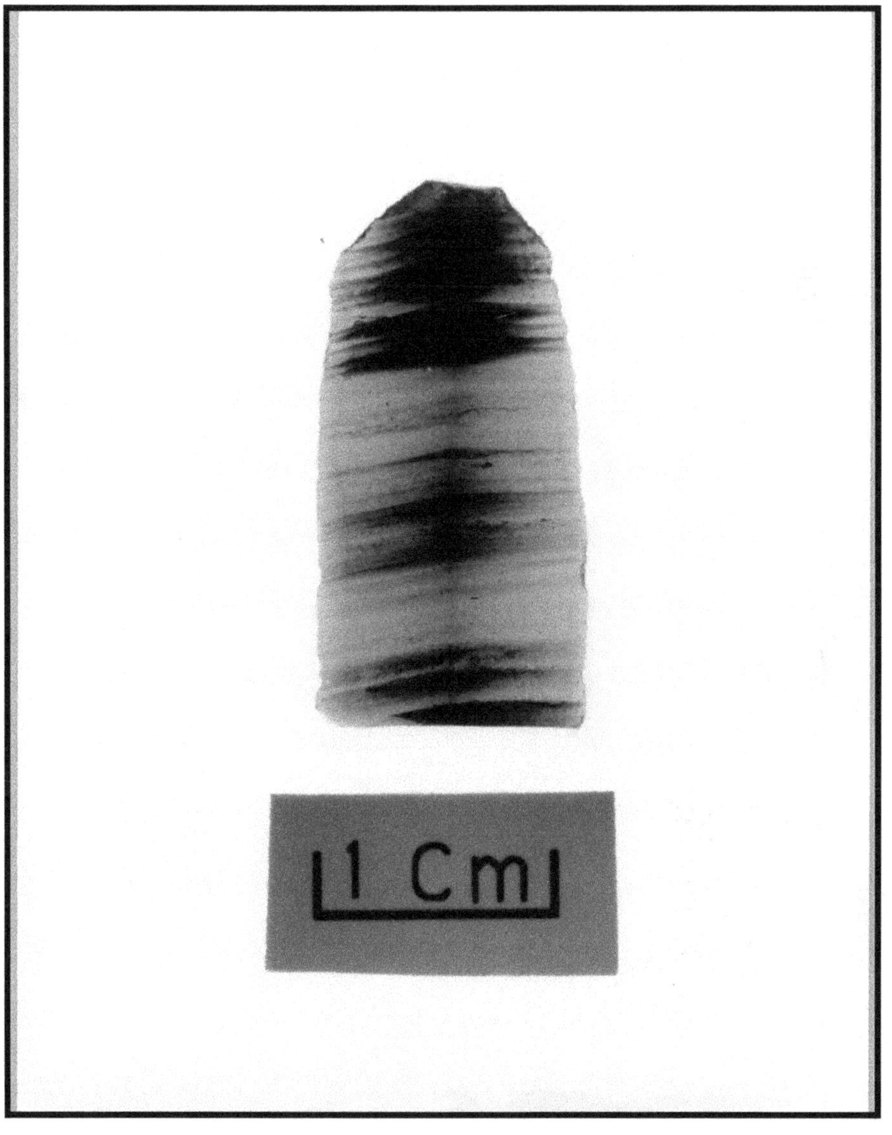


Photograph of "Bands" in Obsidian

Figure I.12

**APPENDIX II**
**OBSIDIAN FROM INDIVIDUAL SITES GROUPED BY GEOGRAPHIC REGION**

Obsidian from Coastal Sites

Table II.1

| Site | Date | Source | Quantity | Archaeological Context | Social Context | Functional Context |
|---|---|---|---|---|---|---|
| **Cerros** | | | | | | |
| | LF | ELC | 16 | CF | IND | Waste |
| | EC | ELC | 1 | CF | IND | Waste |
| | EC | IXT | 1 | CF | IND | Waste |
| | TC | ELC | 1 | CF | IND | Waste |
| | | | | | | |
| **Moho Cay** | | | | | | |
| | EC | IXT | 1 | M | NE1 | Waste |
| | EC | ELC | 4 | M | NE1 | Waste |
| | EC | ELC | 4 | MG | NE1 | Burial |
| | EC | ELC | 2 | M | NE1 | Waste |
| | EC | ELC | 2 | CF | NE1 | Waste |
| | | | | | | |
| **NRL** | | | | | | |
| | LC | ELC | 17 | M | E1 | Waste |
| | LC | IXT | 3 | M | E1 | Waste |

Obsidian from River Sites

Table II.2

| | Date | Source | Quantity | Archaeological Context | Social Context | Functional Context |
|---|---|---|---|---|---|---|
| **Cuello** | | | | | | |
| | LF | ELC | 16 | CF | IND | Waste |
| | LF | IXT | 26 | CF | IND | Waste |
| | LF | OTH | 13 | CF | IND | Waste |
| | EC | ELC | 9 | CF | IND | Waste |
| | EC | IXT | 6 | CF | IND | Waste |
| **Colha** | | | | | | |
| | LF | SMJ | 1 | M | IND | Waste |
| | LF | IXT | 3 | WS | IND | Waste |
| | LF | IXT | 1 | M | IND | Waste |
| | LC | IXT | 1 | M | IND | Waste |
| | LC | ELC | 5 | M | E1 | Waste |
| | LC | IXT | 2 | M | E1 | Waste |
| | LC | ELC | 7 | WS | WS/NE2 | Waste |
| | LC | IXT | 8 | WS | WS/NE2 | Waste |
| | LC | ELC | 2 | WS | WS | Waste |
| | LC | ELC | 2 | M | NE2 | Waste |
| | LC | IXT | 6 | M | NE2 | Waste |
| | LC | IXT | 12 | FF | E1 | Domestic |
| | LC | SMJ | 1 | FF | E1 | Domestic |
| | LC/TC | ELC | 21 | MG | RC2 | Burial |
| | LC/TC | IXT | 12 | MG | RC2 | Burial |
| | LC/TC | ELC | 4 | WS | WS/NE2 | Waste |

| | | | | | | |
|---|---|---|---|---|---|---|
| | TC | ELC | 5 | WS | WS | Waste |
| | TC | IXT | 12 | WS | WS | Waste |
| | LC | ELC | 8 | UNK | NE2 | Indeterminate |
| | LC | IXT | 5 | UNK | NE2 | Indeterminate |
| | LC | ELC | 3 | UNK | E2 | Indeterminate |
| | LC | IXT | 17 | UNK | E2 | Indeterminate |
| | LC/TC | ELC | 26 | UNK | E2 | Indeterminate |
| | LC/TC | IXT | 16 | UNK | E2 | Indeterminate |
| | TC | ELC | 3 | UNK | E1 | Indeterminate |
| | TC | IXT | 1 | UNK | E1 | Indeterminate |
| | TC | SMJ | 1 | UNK | E1 | Indeterminate |
| **Nohmul** | | | | | | |
| | LF | ELC | 1 | CF | IND | Waste |
| | EC | ELC | 1 | CF | IND | Waste |
| | LC | ELC | 8 | CF | IND | Waste |
| | LC | IXT | 19 | CF | IND | Waste |
| | TC | ELC | 4 | CF | IND | Waste |
| | TC | IXT | 16 | CF | IND | Waste |

Obsidian from Inland Sites

Table II.3

| | **Date** | **Source** | **Quantity** | **Archaeological Context** | **Social Context** | **Functional Context** |
|---|---|---|---|---|---|---|
| **Ka'Kabish** | | | | | | |
| | LC | ELC | 86 | TS | RC2 | Ritual |
| **Kichpanha** | | | | | | |
| | MF | ELC | 1 | CF | IND | Waste |
| | ***MF*** | SMJ | 1 | CF | IND | Waste |
| | MF | IXT | 2 | CF | IND | Waste |
| | LF | ELC | 1 | CF | IND | Waste |
| | LF | IXT | 1 | CF | IND | Waste |
| | LC | ELC | 1 | CF | IND | Waste |
| **Becan** | | | | | | |
| | EC | ELC | 18 | CF | IND | Waste |
| | EC | IXT | 2 | CF | IND | Waste |
| | EC | MX | 2 | CF | IND | Waste |
| | EC | ELC | 51 | C | RC1 | Ritual |
| | EC | MX | 3 | C | RC1 | Ritual |

Obsidian from Central Peten Sites

Table II.4

| | Date | Source | Quantity | Archaeological Context | Social Context | Functional Context |
|---|---|---|---|---|---|---|
| **CPL** | | | | | | |
| | MF | ELC | 11 | CF | IND | Waste |
| | MF | SMJ | 47 | CF | IND | Waste |
| | MF | IXT | 3 | CF | IND | Waste |
| | MF | OTH | 2 | CF | IND | Waste |
| | LF | ELC | 8 | CF | IND | Waste |
| | LF | SMJ | 22 | CF | IND | Waste |
| | LF | IXT | 4 | CF | IND | Waste |
| | EC | ELC | 18 | CF | IND | Waste |
| | EC | SMJ | 6 | CF | IND | Waste |
| | EC | OTH | 1 | CF | IND | Waste |
| | LC | ELC | 40 | CF | IND | Waste |
| | LC | SMJ | 13 | CF | IND | Waste |
| | LC | IXT | 4 | CF | IND | Waste |
| | LC | OTH | 3 | CF | IND | Waste |
| | TC | ELC | 12 | CF | IND | Waste |
| | TC | SMJ | 1 | CF | IND | Waste |
| | TC | OTH | 3 | CF | IND | Waste |
| **El Mirador** | | | | | | |
| | LF | ELC | 6 | CF | IND | Waste |
| | LF | ELC | 1 | COL | IND | Waste |
| | LF | ELC | 1 | FF | IND | Domestic |
| | LF | SMJ | 3 | FF | IND | Domestic |
| | LF | ELC | 3 | FS | IND | Domestic |
| | LF | ELC | 4 | H | IND | Waste |
| | LF | SMJ | 1 | H | IND | Waste |
| | LF | ELC | 3 | M | IND | Waste |
| | LF | ELC | 1 | WS | WS | Waste |
| | LF | ELC | 3 | FS | E1/2 | Domestic |
| | EC | MX | 5 | CF | IND | Waste |
| | EC | MX | 3 | COL | IND | Waste |
| | EC | MX | 1 | FS | E1 | |
| | EC | MX | 4 | H | IND | Waste |
| | LC | ELC | 1 | COL | IND | Waste |
| | LC | ELC | 3 | FF | IND | Domestic |
| | LC | MX | 1 | H | IND | Waste |
| **Tikal/Yaxha** | | | | | | |
| | LF | ELC | 3 | M | NE1 | Waste |
| | LF | SMJ | 10 | M | NE1 | Waste |
| | EC | ELC | 1 | M | NE1 | Waste |
| | EC | SMJ | 2 | M | NE1 | Waste |
| | EC | MX | 1 | M | NE1 | Waste |
| | LC | ELC | 34 | M | NE1 | Waste |
| | LC | SMJ | 3 | M | NE1 | Waste |
| | TC | IXT | 2 | M | NE1 | Waste |
| **Tikal** | | | | | | |
| | LF | ELC | 1 | MG | IND | Burial |
| | LF | SMJ | 4 | CF | IND | Waste |
| | LF | SMJ | 1 | CF | IND | Waste |

| | | | | | | |
|---|---|---|---|---|---|---|
| | LF | ELC | 1 | CF | IND | Waste |
| | EC | MX | 3 | MG | IND | Burial |
| | EC | ELC | 2 | MG | IND | Burial |
| | EC | ELC | 3 | C | RC1 | Ritual |
| | EC | ELC | 1 | M | IND | Waste |
| | EC | MX | 1 | CF | IND | Waste |
| | EC | OTH | 2 | MG | IND | Burial |
| | EC/LC | MX | 3 | CF | IND | Waste |
| | LC | ELC | 5 | TS | RC2 | Ritual |
| | LC | ELC | 1 | C | RC1 | Ritual |
| | TC | ELC | 1 | M | IND | Waste |

Obsidian from the Blue CreekRuin

Table II.5

| Date | Source | Quantity | Archaeological Context | Social Context | Functional Context |
|---|---|---|---|---|---|
| | | | | | |
| LF | ELC | 326 | C | RC1 | Ritual |
| LF | ELC | 7 | CF | IND | Waste |
| LF | ELC | 1 | FF | IND | Domestic |
| LF | ELC | 6 | FS | NE2 5, IND 1 | |
| LF | ELC | 31 | M | NE1 | Waste |
| LF | ELC | 2 | MG | NE1 | Burial |
| LF | ELC | 2 | OTH | IND | Indeterminate |
| LF | ELC | 1 | PZ/S | IND | |
| LF | IXT | 1 | FF | IND | Domestic |
| LF | IXT | 1 | FS | NE2 | Domestic |
| LF | SMJ | 1 | CF | IND | Waste |
| LF | SMJ | 1 | FF | IND | Domestic |
| LF | SMJ | 22 | M | NE1 | Waste |
| LF | SMJ | 1 | PZ/S | NE2 | |
| LF | NA | 146 | C | RC1 145, NE2 1 | Ritual |
| LF | NA | 11 | CF | NE2 4, IND 7 | Waste |
| LF | NA | 1 | FF | | Domestic |
| LF | NA | 1 | FS | | Domestic |
| LF | NA | 61 | M | NE1 60, NE2 1 | Waste |
| LF | NA | 3 | MG | NE1 1, NE3 2 | Burial |
| LF | NA | 4 | PZ/S | NE2 | Indeterminate |
| EC | ELC | 135 | C | E1 134, RC1 1 | Ritual |
| EC | ELC | 7 | CF | IND | Waste |
| EC | ELC | 1 | COL | IND | Waste |
| EC | ELC | 7 | FF | E1 2, NE1 1, IND 4 | Domestic |
| EC | ELC | 2 | MG | E1 | Burial |
| EC | IXT | 1 | FF | IND | Domestic |
| EC | IXT | 1 | FS | IND | Domestic |
| EC | IXT | 1 | C | RC1 | Ritual |
| EC | SMJ | 1 | C | E1 | Ritual |

| | | | | | |
|---|---|---|---|---|---|
| EC | SMJ | 1 | CF | IND | Waste |
| EC | NA | 106 | C | E1 103, RC1 3 | Ritual |
| EC | NA | 2 | CF | IND | Waste |
| EC | NA | 2 | FF | E1 1, IND 1 | Domestic |
| EC | NA | 1 | OTH | IND | Indeterminate |
| LC | ELC | 15 | C | E2 | Ritual |
| LC | ELC | 6 | CF | IND | Waste |
| LC | ELC | 1 | COL | IND | Waste |
| LC | ELC | 1 | FF | IND | Domestic |
| LC | ELC | 13 | FS | E1 | Domestic |
| LC | ELC | 4 | H | IND | Waste |
| LC | ELC | 5 | H/COL | IND | Waste |
| LC | ELC | 1 | H/CF | IND | Waste |
| LC | ELC | 2 | H/FS | E1 | Waste |
| LC | ELC | 2 | OTH | IND | Indeterminate |
| LC | ELC | 1 | PZ/S | NE2 | Indeterminate |
| LC | ELC | 47 | SD | E1 16,<br>E2 25,<br>RC2 6 | Ritual |
| LC | IXT | 1 | H/CF | IND | Waste |
| LC | IXT | 1 | OTH | IND | Indeterminate |
| LC | IXT | 1 | SD | RC2 | Ritual |
| LC | IXT | 1 | CF | IND | Waste |
| LC | SMJ | 2 | H/CF | IND | Waste |
| LC | SMJ | 1 | OTH | E1 | Indeterminate |
| LC | MX | 1 | SD | E2 | Ritual |
| LC | NA | 6 | C | E2 | Ritual |
| LC | NA | 1 | FS | NE2 | Domestic |
| LC | NA | 3 | H | E1 1, IND 2 | Waste |
| LC | NA | 1 | H/COL | IND | Waste |
| LC | NA | 1 | H/FS | E1 | Waste |
| LC | NA | 2 | OTH | IND | Indeterminate |
| LC | NA | 25 | SD | E1 4,<br>E2 15,<br>RC2 5 | Ritual |

## APPENDIX III

### RESULTS OF 91 NAA TEXTED ARTEFACTS VISUALLY IDENTIFIED FROM BLUE CREEK RUIN, ORANGE WALK DISTRICT, BELIZE

| Artefact Number | Assigned Source | Actual Source |
|---|---|---|
| 224 | ELC | ELC |
| 1351 | ELC | ELC |
| 1360 | n/a | ELC |
| 1372 | ELC | ELC |
| 1386 | ELC | ELC |
| 1500 | ELC | ELC |
| 1588 | ELC | ELC |
| 1596 | ELC | ELC |
| 1601 | ELC | ELC |
| 1698 | n/a | ELC |
| 1787 | ELC | ELC |
| 2321 | ELC | ELC |
| 2362 | n/a | ELC |
| 2363 | n/a | ELC |
| 2364 | ELC | ELC |
| 2434 | ELC | ELC |
| 2466 | n/a | IXT |
| 2479 | ELC | ELC |
| 2504 | n/a | ELC |
| 2538 | ELC | ELC |
| 2547 | n/a | SMJ-1 |
| 2601 | ELC | ELC |
| 2603 | n/a | ELC |
| 2604 | n/a | ELC |
| 2617 | IXT | IXT |
| 2628 | ELC | ELC |
| 2630 | ELC | ELC |
| 2637 | ELC | ELC |
| 2667 | ELC | ELC |
| 2777 | n/a | ELC |
| 2803 | ELC | ELC |
| 2804 | n/a | SMJ-1 |
| 2839 | ELC | ELC |
| 2850 | ELC | ELC |
| 2856 | ELC | SMJ-1 |
| 2967 | ELC | ELC |
| 3295 | n/a | ELC |
| 3297 | n/a | ELC |
| 3300 | n/a | ELC |
| 3360 | n/a | ELC |
| 3363 | n/a | ELC |
| 3419 | IXT? | IXT |
| 3420 | IXT? | ELC |
| 3422 | n/a | IXT |
| 3428 | n/a | SMJ-1 |
| 3435 | n/a | ELC |

| | | |
|---|---|---|
| 3448 | n/a | ELC |
| 3449 | n/a | ELC |
| 3451 | ELC | ELC |
| 3497 | ELC | ELC |
| 3503 | ELC | ELC |
| 3505 | ELC | ELC |
| 3512 | ELC | ELC |
| 3521 | ELC | ELC |
| 3524 | ELC | ELC |
| 3530 | ELC | ELC |
| 3538 | ELC | ELC |
| 3542 | n/a | ELC |
| 3557 | ELC | ELC |
| 3561 | n/a | SMJ-1 |
| 3562 | n/a | SMJ-1 |
| 3570 | n/a | ELC |
| 3575 | ELC | ELC |
| 3576 | n/a | ELC |
| 3629 | ELC | ELC |
| 3646 | IXT | IXT |
| 3662 | n/a | ELC |
| 3668 | ELC | ELC |
| 3692 | n/a* noted as 'highly unusual' | ZAR |
| 3737 | ELC | ELC |
| 3739 | n/a | ELC |
| 3742 | ELC | ELC |
| 3762 | n/a | ELC |
| 3766 | ELC | ELC |
| 3794 | n/a | ELC |
| 3872 | ELC | ELC |
| 3874 | ELC | ELC |
| 3877 | n/a | ELC |
| 3878 | ELC | ELC |
| 3927 | ELC | ELC |
| 3952 | n/a | ELC |
| 3973 | ELC | ELC |
| 3983 | ELC | ELC |
| 3992 | ELC | ELC |
| 4033 | n/a | SMJ-1 |
| 4083 | n/a | ELC |
| 4129 | ELC | ELC |
| 4131 | ELC | ELC |
| 4175 | ELC | ELC |
| 4178 | n/a | ELC |
| 4183 | n/a | IXT |

## APPENDIX IV

### ELEMENTAL COMPOSITION OF ARTEFACTS FROM THE BLUE CREEK RUIN TESTED BY NAA (MURR RESULTS)

| Artefact | Ba | Cl | Dy | K | Mn | Na | Source |
|---|---|---|---|---|---|---|---|
| BC# | (ppm) | (ppm) | (ppm) | (ppm) | (ppm or %) | (ppm or %) | |
| 15 | 921.6 | 595.4 | 2.417 | 37574.2 | 657.7 | 30548.0 | ELC |
| 17 | 1037.2 | 587.6 | 2.179 | 33850.1 | 542.8 | 28586.3 | SMJ |
| 19 | 1070.1 | 643.3 | 2.354 | 34910.3 | 470.5 | 29485.3 | IXT |
| 44 | 834.9 | 638.9 | 2.883 | 32710.2 | 658.5 | 30505.1 | ELC |
| 45 | 862.0 | 632.7 | 2.891 | 34120.3 | 670.0 | 30974.8 | ELC |
| 53 | 896.8 | 627.4 | 2.090 | 34520.3 | 633.0 | 30769.4 | ELC |
| 60 | 804.0 | 735.1 | 2.248 | 35694.6 | 660.8 | 30591.2 | ELC |
| 67 | 844.8 | 967.0 | 2.122 | 35491.5 | 666.0 | 30964.8 | ELC |
| 71 | 906.9 | 823.5 | 2.851 | 33429.3 | 653.9 | 30217.0 | ELC |
| 79 | 825.8 | 861.7 | 2.764 | 34375.7 | 652.4 | 30297.3 | ELC |
| 90 | 907.9 | 905.6 | 2.609 | 33690.3 | 664.5 | 30846.4 | ELC |
| 93 | 912.2 | 886.1 | 3.060 | 34783.8 | 657.8 | 30462.3 | ELC |
| 133 | 848.4 | 758.3 | 2.753 | 35218.6 | 659.0 | 30680.9 | ELC |
| 137 | 896.1 | 709.7 | 2.259 | 34239.0 | 656.9 | 30479.6 | ELC |
| 139 | 936.8 | 735.5 | 2.907 | 31357.5 | 661.5 | 30687.6 | ELC |
| 141 | 778.9 | 738.1 | 2.512 | 34440.5 | 667.9 | 30719.6 | ELC |
| 142 | 838.5 | 667.4 | 2.570 | 32321.3 | 656.3 | 30108.2 | ELC |
| 143 | 894.6 | 800.6 | 2.730 | 32576.1 | 664.5 | 30750.5 | ELC |
| 144 | 894.4 | 673.9 | 2.606 | 37028.9 | 666.4 | 30897.5 | ELC |
| 145 | 858.2 | 695.0 | 2.406 | 34416.6 | 658.8 | 30452.9 | ELC |
| 147 | 889.9 | 579.7 | 2.335 | 35326.8 | 655.4 | 30436.8 | ELC |
| 148 | 843.5 | 684.7 | 2.079 | 33782.4 | 662.5 | 30585.9 | ELC |
| 149 | 894.7 | 757.7 | 2.282 | 32476.3 | 662.5 | 30967.9 | ELC |
| 150 | 909.9 | 667.6 | 2.538 | 33126.2 | 650.2 | 30061.1 | ELC |
| 151 | 862.4 | 669.8 | 2.626 | 32389.8 | 643.2 | 30215.3 | ELC |
| 153 | 822.3 | 642.4 | 2.037 | 28706.6 | 645.4 | 30317.0 | ELC |
| 155 | 1158.7 | 762.0 | 2.004 | 31908.3 | 539.3 | 28662.7 | SMJ |
| 156 | 865.1 | 663.7 | 2.928 | 31674.1 | 645.1 | 30539.6 | ELC |
| 157 | 943.7 | 769.3 | 2.325 | 32307.7 | 456.3 | 29230.5 | IXT |
| 158 | 798.0 | 720.0 | 2.437 | 30711.6 | 641.2 | 30375.9 | ELC |
| 159 | 1171.4 | 719.5 | 1.895 | 32889.6 | 455.6 | 29262.9 | IXT |
| 160 | 924.4 | 656.7 | 2.616 | 33423.2 | 641.1 | 30427.4 | ELC |
| 161 | 877.0 | 548.1 | 2.442 | 30372.7 | 647.4 | 30362.6 | ELC |
| 162 | 1012.9 | 616.9 | 1.929 | 33670.1 | 531.9 | 28431.5 | SMJ |
| 165 | 904.5 | 552.6 | 2.176 | 34063.8 | 644.0 | 30230.8 | ELC |

| 185 | 1083.3 | 642.5 | 2.041 | 30894.3 | 512.4 | 28415.0 | SMJ |
|---|---|---|---|---|---|---|---|
| 186 | 1019.5 | 746.7 | 1.760 | 33813.1 | 450.9 | 29079.5 | IXT |
| 191 | 956.0 | 640.9 | 3.298 | 32393.6 | 652.9 | 30668.8 | ELC |
| 202 | 770.1 | 557.6 | 2.859 | 28520.0 | 651.4 | 30692.7 | ELC |
| 206 | 814.3 | 658.5 | 2.803 | 32944.4 | 659.3 | 30749.5 | ELC |
| 207 | 827.8 | 618.2 | 2.144 | 31605.7 | 647.7 | 30409.8 | ELC |
| 209 | 882.3 | 706.4 | 2.491 | 33472.6 | 645.8 | 30483.7 | ELC |
| 212 | 844.2 | 587.9 | 2.824 | 35520.2 | 646.7 | 30495.6 | ELC |
| 219 | 934.7 | 745.5 | 2.227 | 33396.3 | 460.1 | 29567.1 | IXT |
| 224 | 813 | 501 | 2.16 | 3.40% | 649 | 3.07% | ELC |
| 237 | 313.0 | 578.2 | 3.000 | 33195.8 | 672.3 | 30643.6 | ELC |
| 238 | 916.5 | 569.8 | 2.313 | 42057.7 | 474.1 | 26709.3 | IXT |
| 241 | 727.8 | 812.3 | 2.278 | 31442.5 | 645.4 | 30436.6 | ELC |
| 243 | 851.2 | 671.2 | 2.424 | 32834.6 | 638.8 | 30081.7 | ELC |
| 248 | 815.0 | 676.5 | 2.988 | 2770.1 | 640.0 | 30386.2 | ELC |
| 1339 | 789.8 | 674.5 | 2.414 | 31419.5 | 638.5 | 29972.2 | ELC |
| 1340 | 875.8 | 675.9 | 2.422 | 30543.3 | 643.6 | 30320.4 | ELC |
| 1351 | 842 | 519 | 2.07 | 3.57% | 651 | 3.03% | ELC |
| 1360 | 958 | 449 | 2.37 | 3.38% | 664 | 3.07% | ELC |
| 1372 | 905 | 533 | 2.47 | 3.29% | 647 | 3.04% | ELC |
| 1380 | 1033 | 515 | 2.39 | 3.44% | 456 | 2.94% | IXT |
| 1386 | 857 | 513 | 2.61 | 3.40% | 655 | 3.08% | ELC |
| 1390 | 936.8 | 554.8 | 2.211 | 34641.7 | 649.1 | 30529.1 | ELC |
| 1409 | 895 | 555 | 2.70 | 3.45% | 621 | 2.98% | ELC |
| 1421 | 886.1 | 513.3 | 2.325 | 32815.7 | 647.2 | 30787.1 | ELC |
| 1426 | 1169 | 524 | 2.03 | 3.21% | 537 | 2.88% | SMJ |
| 1433 | 1009 | 413.5 | 1.754 | 34555.8 | 506.1 | 27654.1 | SMJ |
| 1434 | 1351.6 | 546.5 | 1.822 | 37097.1 | 527.7 | 28408.6 | SMJ |
| 1437 | 1110 | 517 | 2.26 | 3.03% | 533 | 2.87% | SMJ |
| 1438 | 1060 | 498 | 3.19 | 3.39% | 667 | 3.11% | ELC |
| 1445 | 1089 | 544 | 1.91 | 3.09% | 544 | 2.91% | SMJ |
| 1447 | 1033 | 520 | 1.90 | 3.17% | 532 | 2.87% | SMJ |
| 1450 | 1191 | 557 | 1.96 | 3.24% | 536 | 2.88% | SMJ |
| 1451 | 1116.6 | 576.2 | 1.881 | 33195.4 | 534.5 | 28716.4 | SMJ |
| 1453 | 929.3 | 436.3 | 1.966 | 34101.6 | 646.8 | 30546.4 | ELC |
| 1455 | 1020 | 524 | 1.66 | 3.34% | 537 | 2.86% | SMJ |
| 1456 | 957 | 535 | 3.09 | 3.33% | 644 | 3.02% | ELC |
| 1458 | 1121 | 521 | 1.86 | 3.15% | 524 | 2.80% | SMJ |
| 1459 | 1042 | 569 | 2.50 | 3.17% | 548 | 2.95% | SMJ |
| 1460 | 957.7 | 508.5 | 2.181 | 35566.2 | 534.1 | 28859.4 | SMJ |
| 1463 | 904 | 529 | 2.45 | 3.25% | 649 | 3.07% | ELC |

| | | | | | | | |
|---|---|---|---|---|---|---|---|
| 1464 | 870.3 | 474.9 | 2.401 | 961.1 | 648.8 | 30504.7 | ELC |
| 1500 | 695 | 462 | 2.08 | 3.18% | 648 | 3.04% | ELC |
| 1536 | 741 | 486.6 | 2.362 | 33170.3 | 639.7 | 30293.3 | ELC |
| 1537 | 808 | 454.4 | 1.915 | 36946.3 | 650.8 | 30894.8 | ELC |
| 1539 | 916.9 | 422.9 | 2.650 | 37685.1 | 636.7 | 30120.8 | ELC |
| 1544 | 1013.8 | 523.4 | 1.676 | 38829.9 | 640.5 | 30325.5 | ELC |
| 1547 | 963.4 | 522.7 | 2.242 | 34606.3 | 659.7 | 31230.6 | ELC |
| 1550 | 946.5 | 502.5 | 2.891 | 36679.9 | 655.9 | 30945.0 | ELC |
| 1552 | 1005.1 | 405.7 | 2.391 | 37140.9 | 639.4 | 30282.1 | ELC |
| 1559 | 873.1 | 460.2 | 2.427 | 32772.5 | 637.2 | 30026.5 | ELC |
| 1564 | 933.9 | 454.5 | 2.004 | 31302.1 | 635.6 | 30101.3 | ELC |
| 1588 | 1000 | 493 | 2.07 | 3.18% | 655 | 3.04% | ELC |
| 1596 | 828 | 513 | 2.55 | 3.50% | 651 | 3.03% | ELC |
| 1601 | 784 | 458 | 2.4 | 3.15% | 650 | 3.00% | ELC |
| 1698 | 875 | 461 | 2.26 | 3.17% | 642 | 3.02% | ELC |
| 1787 | 804 | 522 | 2.63 | 3.37% | 634 | 2.99% | ELC |
| 1891 | 988.3 | 436.5 | 2.112 | 32080.7 | 635.4 | 30177.1 | ELC |
| 1928 | 956 | 418 | 2.41 | 3.27% | 646 | 3.05% | ELC |
| 2233 | 1030.7 | 393.3 | 2.728 | 32343.1 | 637.0 | 30086.6 | ELC |
| 2245 | 1090 | 638 | 2.22 | 3.14% | 529 | 2.86% | ELC |
| 2251 | 1020 | 495 | 3.27 | 3.29% | 648 | 3.09% | ELC |
| 2252 | 1030 | 501 | 1.88 | 3.35% | 536 | 2.86% | SMJ |
| 2256 | 879 | 438.9 | 2.007 | 30961.1 | 639.9 | 30306.6 | ELC |
| 2261 | 1127 | 511 | 1.82 | 3.24% | 544 | 2.92% | SMJ |
| 2276 | 1061.9 | 494.7 | 2.066 | 35586.8 | 642.1 | 30428.2 | ELC |
| 2283 | 1012 | 465 | 3.04 | 3.62% | 525 | 2.84% | SMJ |
| 2291 | 1065.7 | 434.2 | 1.826 | 33235.2 | 538.6 | 28850.3 | SMJ |
| 2292 | 1113 | 534 | 2.02 | 3.36% | 542 | 2.91% | SMJ |
| 2321 | 848 | 543 | 2.53 | 3.34% | 649 | 3.04% | ELC |
| 2322 | 862 | 553 | 2.27 | 3.60% | 640 | 3.00% | ELC |
| 2362 | 928 | 484 | 2.32 | 3.61% | 650 | 3.05% | ELC |
| 2363 | 1064 | 465 | 2.44 | 3.50% | 698 | 3.23% | ELC |
| 2364 | 808 | 502 | 2.04 | 3.19% | 663 | 3.09% | ELC |
| 2369 | 1038.8 | 487.6 | 1.728 | 33365.9 | 537.1 | 28938.7 | SMJ |
| 2376 | 1198.1 | 480.2 | 1.740 | 31058.9 | 533.3 | 28583.0 | SMJ |
| 2379 | 1201.4 | 479.9 | 1.927 | 31596.5 | 527.4 | 28371.5 | SMJ |
| 2434 | 755 | 437 | 2.43 | 3.36% | 655 | 3.02% | ELC |
| 2466 | 975 | 498 | 2.66 | 3.59% | 459 | 2.93% | IXT |
| 2479 | 830 | 494 | 2.42 | 3.34% | 654 | 3.05% | ELC |
| 2504 | 818 | 587 | 2.72 | 3.52% | 654 | 3.05% | ELC |
| 2538 | 751 | 484 | 2.22 | 3.23% | 653 | 3.11% | ELC |

| 2547 | 920 | 543 | 1.82 | 3.51% | 538 | 2.86% | SMJ |
|---|---|---|---|---|---|---|---|
| 2601 | 798 | 446 | 2.67 | 3.12% | 660 | 3.06% | ELC |
| 2603 | 831 | 523 | 2.71 | 3.38% | 653 | 3.11% | ELC |
| 2604 | 833 | 575 | 1.81 | 3.49% | 650 | 3.02% | ELC |
| 2617 | 935 | 568 | 2.46 | 3.24% | 458 | 2.92% | IXT |
| 2628 | 810 | 464 | 2.37 | 3.50% | 668 | 3.10% | ELC |
| 2630 | 866 | 528 | 3.02 | 3.48% | 652 | 3.04% | ELC |
| 2637 | 760 | 504 | 2.61 | 3.57% | 658 | 3.07% | ELC |
| 2667 | 836 | 436 | 2.33 | 3.29% | 661 | 3.07% | ELC |
| 2777 | 818 | 448 | 2.23 | 3.15% | 646 | 3.05% | ELC |
| 2803 | 789 | 495 | 1.29 | 3.51% | 651 | 3.06% | ELC |
| 2804 | 923 | 459 | 1.91 | 3.15% | 549 | 2.87% | SMJ |
| 2839 | 810 | 548 | 2.43 | 3.24% | 663 | 3.08% | ELC |
| 2850 | 910 | 539 | 2.45 | 3.07% | 655 | 3.06% | ELC |
| 2856 | 1012 | 403 | 1.98 | 3.27% | 525 | 2.81% | SMJ |
| 2967 | 844 | 434 | 2.73 | 3.60% | 645 | 3.04% | ELC |
| 3295 | 921 | 504 | 2.39 | 3.02% | 634 | 3.02% | ELC |
| 3297 | 884 | 568 | 3.03 | 3.24% | 654 | 3.09% | ELC |
| 3300 | 887 | 472 | 2.55 | 3.20% | 642 | 3.03% | ELC |
| 3360 | 948 | 502 | 2.34 | 3.31% | 637 | 3.02% | ELC |
| 3363 | 904 | 493 | 2.36 | 3.40% | 637 | 3.03% | ELC |
| 3419 | 956 | 554 | 2.02 | 3.51% | 455 | 2.93% | IXT |
| 3420 | 848 | 591 | 2.32 | 3.12% | 646 | 3.04% | ELC |
| 3422 | 1061 | 575 | 2.81 | 3.26% | 449 | 2.91% | IXT |
| 3428 | 1003 | 487 | 1.95 | 3.60% | 525 | 2.83% | SMJ |
| 3435 | 1016 | 463 | 2.77 | 3.39% | 651 | 3.08% | ELC |
| 3448 | 886 | 507 | 2.44 | 3.24% | 662 | 3.14% | ELC |
| 3449 | 902 | 452 | 2.85 | 3.33% | 652 | 3.07% | ELC |
| 3451 | 897 | 496 | 2.44 | 3.64% | 641 | 2.82% | ELC |
| 3497 | 939 | 504 | 2.52 | 3.17% | 644 | 3.09% | ELC |
| 3503 | 866 | 530 | 2.66 | 3.28% | 642 | 3.04% | ELC |
| 3505 | 1001 | 493 | 2.45 | 3.33% | 643 | 3.05% | ELC |
| 3512 | 849 | 468 | 2.83 | 3.49% | 653 | 3.07% | ELC |
| 3521 | 1033 | 502 | 2.35 | 3.54% | 649 | 3.08% | ELC |
| 3524 | 931 | 549 | 2.67 | 3.10% | 643 | 3.02% | ELC |
| 3530 | 885 | 503 | 2.30 | 3.07% | 655 | 3.11% | ELC |
| 3538 | 944 | 513 | 2.85 | 3.50% | 648 | 3.07% | ELC |
| 3542 | 958 | 536 | 1.98 | 3.45% | 663 | 3.10% | ELC |
| 3557 | 975 | 439 | 2.83 | 3.45% | 656 | 3.09% | ELC |
| 3561 | 1217 | 489 | 2.07 | 3.33% | 539 | 2.89% | SMJ |
| 3562 | 1030 | 478 | 2.51 | 3.34% | 533 | 2.87% | SMJ |

| 3570 | 962 | 521 | 2.77 | 3.51% | 653 | 3.10% | ELC |
|---|---|---|---|---|---|---|---|
| 3575 | 964 | 544 | 3.11 | 3.18% | 648 | 3.05% | ELC |
| 3576 | 924 | 453 | 3.19 | 3.39% | 655 | 3.11% | ELC |
| 3629 | 887 | 478 | 3.04 | 3.33% | 652 | 3.07% | ELC |
| 3646 | 992 | 502 | 2.31 | 3.42% | 458 | 2.94% | IXT |
| 3662 | 1014 | 555 | 1.80 | 3.46% | 659 | 3.10% | ELC |
| 3668 | 1009 | 459 | 2.78 | 3.89% | 644 | 3.05% | ELC |
| 3692 | 439 | 519 | 4.61 | 4.03% | 256 | 2.96% | ZAR |
| 3737 | 805 | 438 | 2.08 | 3.20% | 657 | 3.05% | ELC |
| 3739 | 797 | 563 | 2.07 | 3.34% | 651 | 3.07% | ELC |
| 3742 | 748 | 537 | 2.73 | 3.28% | 658 | 3.04% | ELC |
| 3762 | 877 | 384 | 2.45 | 3.19% | 658 | 3.07% | ELC |
| 3766 | 842 | 449 | 2.74 | 3.19% | 649 | 3.05% | ELC |
| 3794 | 929 | 485 | 2.54 | 3.43% | 655 | 3.06% | ELC |
| 3872 | 789 | 502 | 2.41 | 3.44% | 655 | 3.04% | ELC |
| 3874 | 853 | 509 | 2.29 | 3.38% | 660 | 3.08% | ELC |
| 3877 | 985 | 596 | 2.66 | 3.37% | 653 | 3.09% | ELC |
| 3878 | 926 | 456 | 2.06 | 2.97% | 651 | 3.06% | ELC |
| 3927 | 994 | 435 | 2.28 | 3.36% | 648 | 3.02% | ELC |
| 3952 | 920 | 543 | 1.95 | 3.10% | 655 | 3.07% | ELC |
| 3973 | 901 | 540 | 2.82 | 3.45% | 662 | 3.11% | ELC |
| 3983 | 768 | 477 | 2.81 | 3.46% | 649 | 3.06% | ELC |
| 3992 | 983 | 520 | 2.47 | 3.10% | 650 | 3.06% | ELC |
| 4033 | 1122 | 452 | 2.33 | 3.37% | 538 | 2.89% | SMJ |
| 4083 | 835 | 508 | 2.10 | 3.16% | 639 | 3.04% | ELC |
| 4129 | 693 | 472 | 2.59 | 3.40% | 644 | 2.99% | ELC |
| 4131 | 874 | 550 | 2.51 | 3.43% | 654 | 3.09% | ELC |
| 4175 | 785 | 511 | 1.80 | 3.26% | 650 | 3.04% | ELC |
| 4178 | 803 | 527 | 2.04 | 3.36% | 645 | 3.04% | ELC |
| 4183 | 947 | 488 | 2.31 | 3.67% | 460 | 2.94% | IXT |
| 4279 | 933.2 | 819.9 | 2.650 | 36043.4 | 471.0 | 29582.6 | IXT |
| 4280 | 820.9 | 635.3 | 2.632 | 30672.8 | 630.5 | 30227.4 | ELC |
| 4281 | 774.6 | 571.4 | 2.378 | 31031.7 | 639.9 | 30097.0 | ELC |

## APPENDIX V

### DISTRIBUTION OF OBSIDIAN SOURCES AT VARIOUS SITES FROM MIDDLE FORMATIVE TO LATE POST-CLASSIC

**MIDDLE FORMATIVE**

| SITE | SMJ | ELC | IXT | MEX | OTH | REFERENCE |
|---|---|---|---|---|---|---|
| Central Peten Lakes, Guatemala | 48 | 11 | 3 | | | Dreiss and Brown 1989; Rice 1984 |
| Colha, Belize | 7 | | | | | Dreiss 1986 |
| Edzna, Mexico | 12 | | | | | Nelson et al. 1983 |
| El Balsamo, Guatemala | 18 | 2 | | | | Sidrys and Kimberlin 1979 |
| El Mirador, Guatemala | 3 | | | | | Fowler et al. 1989 |
| Guajilar, Mexico | 135 | 9 | | | | Clark 1988 |
| La Libertad, Mexico | 5705 | 33 | | 5 | 6 | Clark 1988 |
| La Venta, Mexico | 128 | 3 | 4 | 325 | 19 | Nelson et al. 1977 |
| Seibal, Guatemala | 24 | 3 | | | | Nelson et al. 1978 |
| Sta. Marta Rosario, Mexico | 204 | 3 | | 1 | | Clark 1988 |

**LATE FORMATIVE**

| SITE | SMJ | ELC | IXT | MEX | OTH | REFERENCE |
|---|---|---|---|---|---|---|
| Becan, Mexico | 3 | 8 | 2 | | | Rovner 1981; Dreiss and Brown 1989 |
| Central Peten Lakes, Guatemala | 21 | 8 | 4 | | 3 | Rice 1984 |
| Cerros, Belize | | 16 | | | | Nelson 1985 |
| Colha, Belize | 4 | 4 | 6 | | | Dreiss and Brown 1989 |
| Edzna, Mexico | 11 | 28 | 2 | | | Nelson et al. 1983 |
| El Balsamo, Guatemala | 7 | 13 | | | | Sidrys and Kimberlin 1979 |
| El Mirador, Guatemala | 4 | 13 | | | | Fowler et al. 1989 |
| Kakalche, Belize | 1 | 7 | | | | Graham 1994 |
| Seibal, Guatemala | 18 | 5 | | | | Nelson et al. 1978 |
| Tikal, Guatemala | 20 | 9 | 10 | 4 | | Moholy-Nagy et al. 1984 |

**EARLY CLASSIC**

| SITE | SMJ | ELC | IXT | MEX | OTH | REFERENCE |
|---|---|---|---|---|---|---|
| Becan/Chicanna, Mexico | 1 | 83 | 2 | 13 | | Rovner 1981; Dreiss and Brown 1989 |
| Caledonia, Belize | | 28 | 1 | | | Neivens et al. 1983 |
| Central Peten Lakes, Guatemala | 6 | 18 | | 1 | | Rice 1984 |
| Coba, Mexico | | 3 | | | | Nelson et al. 1983 |
| Edzna, Mexico | | 7 | | 2 | | Nelson et al. 1983 |
| Moho Caye, Belize | | 13 | 3 | | | Dreiss 1986; Healy et al. 1984 |
| Nohmul, Bel | | 4 | | | | Hammond et al. 1984 |
| Tikal, Guatemala | 4 | 41 | | 22 | | Moholy-Nagy et al. 1984 |

**LATE CLASSIC**

| SITE | SMJ | ELC | IXT | MEX | OTH | REFERENCE |
|---|---|---|---|---|---|---|
| Barton Ramie, Belize | | 6 | | | | Nelson et al. 1978 |
| Becan/Chicanna, Mexico | | 28 | 4 | | | Rovner 1981; Dreiss and Brown 1989 |
| Central Peten Lakes, Guatemala | 13 | 41 | | 2 | 4 | Rice 1984 |
| Coba, Mexico | | 4 | | | | Nelson et al. 1983 |
| Colha, Belize | | 17 | 22 | | | Dreiss et al. 1993 |
| Edzna, Mexico | 12 | 63 | | 25 | | Nelson 1985 |
| El Pozito, Belize | 3 | 69 | | | | Neivens and Libbey 1976; this report |
| Kakabish, Belize | 1 | 127 | 12 | | | this report |
| Lamanai South, Belize | 20 | 147 | 46 | | 20 | Smith and McField 1996 |
| Lubaantun, Belize | | 21 | | | 1 | Hammond 1976 |
| Nohmul, Belize | | 8 | 19 | | | Hammond et al. 1984 |
| No. Ambergris Caye, Belize | | 75 | 26 | | | McKillop 1995 |
| Palenque, Mexico | 1 | 118 | 2 | 4 | | Johnson 1976 |
| Seibal, Guatemala | | 7 | 1 | | 1 | Nelson et al. 1978 |
| Tikal, Guatemala | 1 | 50 | 3 | 7 | | Moholy-Nagy et al. 1984 |

**TERMINAL CLASSIC**

| SITE | SMJ | ELC | IXT | MEX | OTH | REFERENCE |
|---|---|---|---|---|---|---|
| Aventura, Belize | | 18 | 1 | | | Neivens et al. 1983 |
| Becan/Chicanna, Mexico | | 62 | 13 | 11 | | Rovner 1981; Dreiss and Brown 1989 |
| Central Peten Lakes, Guatemala | 1 | 13 | 4 | | | Rice 1984 |
| Chichen Itza, Mexico | | | | 6 | | Andrews et al. 1989 |
| Coba, Mexico | | 4 | | | | Nelson et al. 1983 |
| Colha, Belize | 2 | 81 | 72 | | | Dreiss et al. 1993 |
| Isla Cerritos, Mexico | | 2 | | 12 | | Andrews et al. 1989 |
| Lamanai South, Belize | 15 | 87 | 37 | | 3 | Smith and McField 1996 |
| Nohmul, Belize | | 4 | 16 | | | Hammond et al. 1984 |
| No. Ambergris Caye, Belize | | 42 | 8 | 5 | | McKillop 1995 |
| No. River Lagoon, Belize | | 17 | 3 | | | Dreiss et al. 1993 |
| Patchchacan, Belize | | 5 | 1 | | | Neivens et al. 1983 |
| Point Placencia, Belize | | 6 | | | | Dreiss and Brown 1989 |
| Seibal, Guatemala | 6 | 12 | 3 | 1 | | Nelson et al. 1978 |
| Tikal, Guatemala | 4 | 1 | 2 | | | Moholy-Nagy et al. 1984 |
| Tipu, Belize | | 25 | 13 | | | Dreiss and Brown 1989 |
| Uxmal, Mexico | | 9 | | 1 | | Nelson et al. 1983 |
| Wild Cane Cay, Belize | | 12 | 15 | | 2 | McKillop et al. 1988 |

**EARLY POST-CLASSIC**

| SITE | SMJ | ELC | IXT | MEX | OTH | REFERENCE |
|---|---|---|---|---|---|---|
| Central Peten Lakes, Guatemala | 20 | 27 | 46 | 3 | | Rice 1984 |
| Colha, Belize | | 1 | 13 | | | Dreiss and Brown 1989 |
| Cozumel, Mexico | | 1 | 2 | 3 | | Nelson 1985 |
| Isla Cerritos, Mexico | | 9 | 1 | 31 | 1 | Andrews et al. 1989 |
| Point Placencia, Belize | | 9 | 3 | 3 | | Dreiss and Brown 1989 |
| Sta. Rita Corozal, Belize | 1 | 3 | 12 | 3 | | Neivens et al. 1983 |
| Tipu, Belize | 9 | 18 | 135 | 2 | | Dreiss and Brown 1989 |
| Wild Cane Cay, Belize | 1 | 6 | 63 | | 1 | McKillop et al. 1988 |

**MIDDLE POST-CLASSIC**

| SITE | SMJ | ELC | IXT | MEX | OTH | REFERENCE |
|---|---|---|---|---|---|---|
| Los Renegados, Belize | | 3 | 276 | | | McKillop 1995 |

**MIDDLE/LATE POST-CLASSIC**

| SITE | SMJ | ELC | IXT | MEX | OTH | REFERENCE |
|---|---|---|---|---|---|---|
| Cancun, Mexico | | | 4 | | | Nelson et al. 1983 |
| Cozumel, Mexico | | 1 | 19 | 1 | | Nelson et al. 1983 |
| Sarteneja, Belize | | | 36 | 3 | | Neivens et al. 1983 |
| Tulum, Mexico | | | 10 | | | Nelson et al. 1983 |

## APPENDIX VI

### TABLES OF WEIGHTS FOR BLUE CREEK OBSIDIAN ARTEFACTS

| | ELC | IXT | SMJ | MX | Total |
|---|---|---|---|---|---|
| **LF** | 434.8 | 1.6 | 9.1 | 0.0 | **445.5** |
| **EC** | 69.2 | 3.8 | 2.2 | 0.0 | **75.2** |
| **LC** | 188.1 | 4.7 | 1.7 | 0.7 | **195.2** |

Total Grams of Each Obsidian Source by Time Period
for Artefacts from the Blue Creek Ruin

Table VI.1

| | ELC | IXT | SMJ | MX | Total |
|---|---|---|---|---|---|
| **LF** | 97.6 | 0.4 | 2.0 | 0.0 | **100.0** |
| **EC** | 92.0 | 5.0 | 3.0 | 0.0 | **100.0** |
| **LC** | 96.4 | 2.4 | 0.9 | 0.3 | **100.0** |

Percentage Weight of Each Source by Time Period
for Artefacts from the Blue Creek Ruin

Table VI.2

| | ELC | IXT | SMJ | MX | Total |
|---|---|---|---|---|---|
| **Ritual** | 404.6 | 0 | 0 | 0 | **404.6** |
| **Waste** | 22.5 | 0 | 0.3 | 0 | **22.8** |
| **Domestic** | 2.3 | 1.6 | 8.8 | 0 | **12.7** |
| **Burial** | 3.2 | 0 | 0 | 0 | **3.2** |
| **Indeterminate** | 2.4 | 0 | 0 | 0 | **2.4** |
| | **435** | **1.6** | **9.1** | **0** | **445.7** |

Weights of Late Foramtive Material from the Blue Creek Ruin
by Functional Contexts and Source for Artefacts

Table VI.3

| | ELC | IXT | SMJ | MX |
|---|---|---|---|---|
| **Ritual** | 90.8 | 0 | 0 | 0 |
| **Waste** | 5.0 | 0 | 0.1 | 0 |
| **Domestic** | 0.5 | 0.3 | 2.2 | 0 |
| **Burial** | 0.7 | 0 | 0 | 0 |
| **Indeterminate** | 0.5 | 0 | 0 | 0 |
| | **97.5** | **0.3** | **2.3** | **0** |

Percentage of Weights for Late Formative Material from the Blue Creek Ruin
by Functional Context and Source

Table VI.4

| | ELC | IXT | SMJ | MX | Total |
|---|---|---|---|---|---|
| **Ritual** | 100 | 0 | 0 | 0 | **100** |
| **Waste** | 98.7 | 0 | 1.3 | 0 | **100** |
| **Domestic** | 18.1 | 12.6 | 69.3 | 0 | **100** |
| **Burial** | 100 | 0 | 0 | 0 | **100** |

Percentage of Weight for Late Formative Material from the Blue Creek Ruin in Each Functional Context by Source

Table VI.5

| | ELC | IXT | SMJ | MX | Total |
|---|---|---|---|---|---|
| **Ritual** | 54.1 | 2.8 | 0.6 | 0 | **57.5** |
| **Waste** | 7.4 | 0 | 1.6 | 0 | **9.0** |
| **Domestic** | 4.1 | 1.0 | 0 | 0 | **5.1** |
| **Burial** | 3.6 | 0 | 0 | 0 | **3.6** |
| **Indeterminate** | 0 | 0 | 0 | 0 | **0** |
| | **69.2** | **3.8** | **2.2** | **0** | **75.2** |

Weights of Early Classic Material from the Blue Creek Ruin by Functional Contexts and Source

Table VI.6

| | ELC | IXT | SMJ | MX |
|---|---|---|---|---|
| **Ritual** | 71.9 | 3.7 | 0.8 | 0 |
| **Waste** | 9.8 | 0 | 2.1 | 0 |
| **Domestic** | 5.4 | 1.3 | 0 | 0 |
| **Burial** | 4.8 | 0 | 0 | 0 |
| **Indeterminate** | 0 | 0 | 0 | 0 |
| | **91.9** | **5** | **2.9** | **0** |

Percentage of Weights for Early Classic Material from the Blue Creek Ruin by Functional Context and Source

Table VI.7

| | ELC | IXT | SMJ | MX | Total |
|---|---|---|---|---|---|
| | 94.1 | 4.9 | 1.0 | 0 | **100** |
| ***Ritual*** | | | | | |
| **Waste** | 82.2 | 0 | 17.8 | 0 | **100** |
| **Domestic** | 80.4 | 19.6 | 0 | 0 | **100** |
| **Burial** | 100 | 0 | 0 | 0 | **100** |

Percentage of Weight for Early Classic Material from the Blue Creek Ruin in Each Functional Context by Source

Table VI.8

| | ELC | IXT | SMJ | MX | Total |
|---|---|---|---|---|---|
| **Ritual** | 119.9 | 1.9 | 0 | 0.7 | **122.5** |
| **Waste** | 9.3 | 1.0 | 0 | 0 | **10.3** |
| **Domestic** | 17.6 | 0 | 0 | 0 | **17.6** |
| **Burial** | 0 | 0 | 0 | 0 | **0** |
| **Indeterminate** | 41.3 | 1.8 | 1.7 | 0 | **44.8** |
| | **188.1** | **4.7** | **1.7** | **0.7** | **195.2** |

Weights of Late Classic Material from the Blue Creek Ruin
by Functional Contexts and Source

Table VI.9

| | ELC | IXT | SMJ | MX |
|---|---|---|---|---|
| **Ritual** | 61.4 | 1.0 | 0 | 0.3 |
| **Waste** | 4.8 | 0.5 | 0 | 0 |
| **Domestic** | 9.0 | 0 | 0 | 0 |
| **Burial** | 0 | 0 | 0 | 0 |
| **Indeterminate** | 21.1 | 0.9 | 0.9 | 0 |
| | **96.3** | **2.4** | **0.9** | **0.3** |

Percentage of Weights for Late Classic Material from the Blue Creek Ruin
by Functional Context and Source

Table VI.10

| | ELC | IXT | SMJ | MX | Total |
|---|---|---|---|---|---|
| **Ritual** | 97.9 | 1.5 | 0 | 0.6 | **100** |
| **Waste** | 90.3 | 9.7 | 0 | 0 | **100** |
| **Domestic** | 100 | 0 | 0 | 0 | **100** |
| **Burial** | 0 | 0 | 0 | 0 | **0** |

Percentage of Weight for Late Classic Material from the Blue Creek Ruin
in Each Functional Context by Source

Table VI.11

| | ELC | IXT | SMJ | MX | NA | Total |
|---|---|---|---|---|---|---|
| **C** | 404.6 | 0 | 0 | 0 | 52 | **456.6** |
| **CF** | 3.1 | 0 | 0.3 | 0 | 16.3 | **19.7** |
| **FF** | 0 | 1.4 | 0.8 | 0 | 0.4 | **2.6** |
| **FS** | 2.3 | 0.2 | 8 | 0 | 0.1 | **10.6** |
| **M** | 19.4 | 0 | 0 | 0 | 18.1 | **37.5** |
| **MG** | 3 | 0 | 0 | 0 | 1.9 | **4.9** |
| **PZ/S** | 0.7 | 0 | 0 | 0 | 0.8 | **1.5** |
| **OTH** | 1.7 | 0 | 0 | 0 | 0 | **1.7** |
| Total | 434.8 | 1.6 | 9.1 | 0 | 89.6 | 535.1 |

Weights of Late Formative Obsidian from the Blue Creek Ruin
by Archaeological Context and Source

Table VI.12

| EC | ELC | IXT | SMJ | MX | NA | Total |
|---|---|---|---|---|---|---|
| **C** | 54.1 | 2.8 | 0.6 | 0 | 232.1 | **289.6** |
| **CF** | 4.1 | 0 | 1.6 | 0 | 0.9 | **6.6** |
| **COL** | 3.3 | 0 | 0 | 0 | 0 | **3.3** |
| **FF** | 4.1 | 0.3 | 0 | 0 | 0.7 | **5.1** |
| **FS** | 0 | 0.7 | 0 | 0 | 0 | **0.7** |
| **MG** | 3.6 | 0 | 0 | 0 | 0 | **3.6** |
| **OTH** | 0 | 0 | 0 | 0 | 0.4 | **0.4** |
| Total | 69.2 | 5.4 | 2.2 | 0 | 234.1 | 310.9 |

Weights of Early Classic Obsidian from the Blue Creek Ruin
by Archaeological Context and Source

Table VI.13

| LC | ELC | IXT | SMJ | MX | NA | Total |
|---|---|---|---|---|---|---|
| **C** | 30.3 | 0 | 0 | 0 | 11.9 | 42.2 |
| **CF** | 8.4 | 1 | 0 | 0 | 0 | **9.4** |
| **COL** | 0.9 | 0 | 0 | 0 | 0 | **0.9** |
| **FF** | 0.8 | 0 | 0 | 0 | 0 | **0.8** |
| **FS** | 16.8 | 0 | 0 | 0 | 0 | **16.8** |
| **H** | 2.8 | 0 | 0 | 0 | 3.4 | **6.2** |
| **H/MX** | 22.1 | 1.4 | 0.7 | 0 | 0 | **24.2** |
| **OTH** | 16 | 0.4 | 1 | 0 | 1.1 | **18.5** |
| **PZ/S** | 0.4 | 0 | 0 | 0 | 0 | **0.4** |
| **SD** | 89.6 | 1.9 | 0 | 0.7 | 33.3 | **125.5** |
| **Total** | **188.1** | **4.7** | **1.7** | **0.7** | **49.7** | **244.9** |

Weights of Late Classic Obsidian from the Blue Creek Ruin
by Archaeological Context and Source

Table VI.14

| | ELC | | IXT | | SMJ | | NA | | MX | |
|---|---|---|---|---|---|---|---|---|---|---|
| | **g** | **no.** | **g** | **no.** | **g** | **no.** | **g** | **no.** | **g** | **no.** |
| **b/frags** | 112.0 | 339 | 1.6 | 2 | 5.9 | 17 | 56.0 | 210 | 0 | 0 |
| **cores** | 317.3 | 26 | 0 | 0 | 0 | 0 | 30.2 | 2 | 0 | 0 |
| **core frags** | 2.1 | 2 | 0 | 0 | 0 | 0 | 2.0 | 2 | | |
| **flakes** | 2.8 | 4 | 0 | 0 | 3.2 | 5 | 1.4 | 6 | 0 | 0 |
| **chunks** | 0.6 | 2 | 0 | 0 | 0 | 0 | 0 | 0 | 0 | 0 |
| **misc** | 0 | 0 | 0 | 0 | 0 | 0 | 0 | 0 | 0 | 0 |
| **Total** | 434.8 | 373 | 1.6 | 2 | 9.1 | 22 | 89.6 | 220 | 0 | 0 |
| **Gross** | **535.1** | **617** | | | | | | | | |

Formative Period Artefacts from the Blue Creek Ruin by Type, Source and Weight

Table VI.15

| | ELC | | IXT | | SMJ | | NA | | MX | |
|---|---|---|---|---|---|---|---|---|---|---|
| | **g** | **no.** | **g** | **no.** | | **no.** | **g** | **no.** | **g** | **no.** |
| **b/frags** | 38.4 | 143 | 3.8 | 3 | | 1 | 19.5 | 80 | 0 | 0 |
| **cores** | 22.3 | 3 | 0 | 0 | 0 | 0 | 212.1 | 20 | 0 | 0 |
| **core frags** | 5.4 | 2 | 0 | 0 | 1.6 | 1 | 13.1 | 5 | | |
| **flakes** | 3.1 | 4 | 0 | 0 | 0 | 0 | 5.3 | 6 | 0 | 0 |
| **chunks** | 0 | 0 | 0 | 0 | 0 | 0 | 0 | 0 | 0 | 0 |
| **misc** | 0 | 0 | 0 | 0 | 0 | 0 | 0 | 0 | 0 | 0 |
| **Total** | 69.2 | 152 | 3.8 | 3 | 2.2 | 2 | 250 | 111 | 0 | 0 |
| **Gross** | **325.2** | **268** | | | | | | | | |

Early Classic Artefacts from the Blue Creek Ruin by Type, Source and Weight

Table VI.16

| | ELC | | IXT | | SMJ | | NA | | MX | |
|---|---|---|---|---|---|---|---|---|---|---|
| | **g** | **no.** | **g** | **no.** | **g** | **no.** | **g** | **no.** | **g** | **no.** |
| **b/frags** | 131.7 | 92 | 3.7 | 3 | 1.7 | 3 | 40.3 | 31 | 0.7 | 1 |
| **cores** | 11.2 | 1 | 0 | 0 | 0 | 0 | 0 | 0 | 0 | 0 |
| **core frags** | 5.9 | 1 | 0 | 0 | 0 | 0 | 0 | 0 | 0 | 0 |
| **flakes** | 0 | 0 | 0 | 0 | 0 | 0 | 0 | 0 | 0 | 0 |
| **chunks** | 0 | 0 | 0 | 0 | 0 | 0 | 0 | 0 | 0 | 0 |
| **uniface** | 39.1 | 3 | 0 | 0 | 0 | 0 | 9.4 | 1 | 0 | 0 |
| **biface** | 0 | 0 | 1.0 | 1 | 0 | 0 | 0 | 0 | 0 | 0 |
| **inlay** | 0.2 | 1 | 0 | 0 | 0 | 0 | 0 | 0 | 0 | 0 |
| **Total** | 188.1 | 98 | 4.7 | 4 | 1.7 | 3 | 49.7 | 32 | 0.7 | 1 |
| **Gross** | **244.9** | **138** | | | | | | | | |

Late Classic Artefacts from the Blue Creek Ruin by Type, Source and Weight

Table VI.17

## REFERENCES


Adams, R. E. W.
1978. "Routes of Communication in Mesoamerica: The Northern Guatemalan Highlands and the Peten," in Mesoamerican Communication Routes and Culture Contacts, vol. 40, Papers of the New World Archaeological Foundation. Edited by T. A. Less and C. Navarrete, pp. 27-35. Provo, Utah: New World Archaeological Foundation.

1991. Prehistoric Mesoamerica, revised edition. Norman: University of Oklahoma Press.
Adams, R. E. W., and R. C. Jones. 1981. Spatial Patterns and Regional Growth Among Classic Maya Cities. American Antiquity 46:301-322.

Andrefsky Jr., W.
1998. Lithics: Macroscopic approches to analysis. Cambridge Manuals in Archaeology. Cambridge, UK: Cambridge University Press.
Andrews, A. P. 1980. Ancient Maya Salt Production and Trade. Tucson: University of Arizona Press.

Andrews, A. P.
1984. Long-Distance Exchange Among the Maya: A comment on Marcus. American Antiquity 49:826-828.

1990. "The Role of Trading Ports in Maya Civilization," in Vision and Revision in Maya Studies. Edited by F. S. Chancy and P. D. Harrison, pp. 159-167. Albuquerque: University of New Mexico Press.
Andrews, A. P., F. Asaro, H. V. Michel, F. H. Stross, and P. C. Rivero. 1989. The Obsidian Trade at Isla Cerritos, Yucatan, Mexico. Journal of Field Archaeology 16:355-363.

Andrews V, E. W., and N. Hammond.
1990. Redefinition of the Swasey Phase at Cuello. American Antiquity 55:570-584.

Arnaud, M. C.
1990. El Comercio Clasico de Obsidiana: Rutas entre Tierras Altas y Terras Bajas en el Area Maya. Latin American Antiquity 1:347-367.

Asaro, F., and H. Michel.
1981. "Obsidian Source Analysis," in Conditions for the Evolution of Complex Societies: Development of the Central Lowland Maya. Edited by A. Ford, pp. 252-259. Santa Barbara, California: Ph.D. Dissertation, University of California.

Asaro, F., H. V. Michel, R. V. Sidrys, and F. Stross.
1978. High Precision Chemical Characterization of Major Obsidian Sources in Guatemala. American Antiquity 43:436-443.

Awe, J., and P. Healy.
1994. Flakes to Blades? Middle Formative Development of Obsidian Artifacts in the Upper Belize River Valley. Latin American Antiquity 5:195-205.

Awe, J. J.
1994. "Ritual, Religion and Cultural Complexity in the Middle Formative Belize River Valley," . The University Museum, Philidelphia: Paper Presented at the Philidelphia Maya Conference.

Awe, J. J., C. Bill, M. Campbell, and D. Cheetham.
1990. Early Middle Formative Occupation in the Central Maya Lowlands: Recent Evidence from Cahal Pech, Belize. Papers from the Institute of Archaeology, UCL 1:1-5.
Bahn, P. 1992. Collins Dictonary of Archaeology. Glasglow: Harper Collins Publisher.

Baker, J.
1997. "Investigations in the Ditched Agricultural Fields," in The Blue Creek Project: Working Papers from the 1996 Field Season. Edited by W. D. Driver, H. L. Clagett, and H. R. Haines, pp. 71-78. San Antonio: Maya Research Program, St. Mary's University.

Baker, R.
1995. "No Postclassic Here: A Reconnaissance of the Lamanai-Kakabish Corridor," in Archaeological Research at Blue Creek, Belize. Progress Report of the Third (1994) Field Season. Edited by T. H. Guderjan and W. D. Driver, pp. 105-118. San Antonio: Maya Research Program, St. Mary's University.

1996. "Status Report on the Settlement Survey," in Archaeological Research at Blue Creek, Belize. Progress Report of the Fourth (1995) Field Season. Edited by T. H. Guderjan, W. D. Driver, and H. R. Haines, pp. 109-116. San Antonio: Maya Research Program, St. Mary's University.

Barrera Vasquez, A.
1995. Diccionario Maya, 3 edition. Mexico, DF: Editorial Porrua, S. A.

Bateson, J. H., and I. H. S. Hall.
1970. Reconnaissance Geochemical and Geological Investigation of the Maya Mountains of Southern British Honduras. Vol. Report No. 16. London: Institute of Geological Sciences, Overseas Division.

Becker, M. J.
1992. "Burials as Caches; Caches as Burials: A New Interpretation of the Meaning of Ritual Deposits Among the Classic Period Lowland Maya," in New Theories on the Ancient Maya. Edited by E. C. Danien and R. J.

Sharer. Philidelphia: The University Museum, University of Pennsylvania Press.

Bohannan, P.
1966. Social Anthropology: Holt, Rinehart and Winston.

Borhegyi, S. F.
1965a. "Archaeological Synthesis of the Guatemalan Highlands," in Archaeology of Southern Mesoamerica: Part One, vol. 2, Handbook of Middle American Indians. Edited by G. R. Willey, pp. 3-58. Austin: University of Texas Press.

1965b. "Settlement Patterns of the Guatemalan Highlands," in Archaeology of Southern Mesoamerica: Part One, vol. 2, Handbook of Middle American Indians. Edited by G. R. Willey, pp. 59-75. Austin: University of Texas Press.

Brady, J. E., J. W. Ball, R. L. Bishop, D. C. Pring, N. Hammond, and R. A. Housley.
1998. The Lowland Maya "Protoclassic": A Reconsideration of its Nature and Significance. Ancient Mesoamerica 9:17-38.

Braswell, G. E.
1996. A Maya Obsidian Source: The Geoarchaeology, Settlement History and Ancient Economy of San Martin Jilotepeque, Guatemala. Ph.D.

Braswell, G. E., E. W. Andres, V, and M. D. Glascock.
1994. The Obsidian Artifacts of Quelpa, El Salvador. Ancient Mesoamerica 5:173-192.

Braswell, G. E., and M. D. Glascock.
1992. A New Obsidian Source in the Highlands of Guatemala. Ancient Mesoamerica 3:47-49.

Brown, K.
1977. "The Valley of Guatemala: a Highland Port of Trade," in Teotihuacan and Kaminaljuyu., Monograph Series on Kaminaljuyu. Edited by W. T. Sanders and J. W. Michels, pp. 205-396. Pittsburg: Penn State University Press.

Brumfiel, E. M.
1994. "Factional competition and political development in the New World: an introduction," in Factional competition and political development in the New World, New Directions in Archaeology. Edited by E. M. Brumfiel and J. W. Fox, pp. 3-14. Cambridge, UK: Cambridge University Press.

Bullard, W. R., Jr.
1960. Maya Settlement Patterns in Northeastern Peten, Guatemala. American Antiquity 25:355-372.

Buttles, P. J.
1992. Small Finds in Context: The Preclassic Artifacts of Colha, Belize. Masters, University of Texas at Austin.

CFANR.
"PROTON INDUCED X-RAY EMISSION TECHNIQUE (PIXE)," , vol. 2000: Center for Fundamental and Applied Neutron Research.

Chapman, A. C.
1957. "Port of Trade Enclaves in Aztec and Maya Civilizations," in Trade and Market in the Early Empires. Edited by K. Polanyi, C. M. Arensberg, and H. Peueson, pp. 144-153. New York: The Free Press.

Chapman, A. M.
1971. "Commentary on: Mesoamerican Trade and its Role in the Emergence of Civilization," in Observations on the Emergence of Civilization in Mesoamerica, vol. 11, Contributions of the University of California Archaeological Research Facility. Edited by R. F. Heizer and J. A. Graham, pp. 196-211. Berkeley: California Archaeological Research Foundation, University of California.

Charlton, T. H.
1969. On the Identificiation of Pre-Hispanic Obsidian Mines in Southern Hidalgo. American Antiquity 34:176-177.

Chase, D. Z., and A. F. Chase.
1989. "Routes of Trade and Communication and the Integration of Maya Society: The Vista From Santa Rita Corozal, Belize," in Coastal Maya Trade, vol. 8, Trent University Occasional Papers in Anthropology. Edited by H. McKillop and P. F. Healy, pp. 19-32. Peterborough: Trent University.

Cheetham, D.
1994. "Ancient Causeways, Elite Ritual and Settlement Patterns of the Maya: Recent Evidence from Cahal Pech, Cayo, Belize," . Anaheim, California: Presented at the 59th Annual Meeting of hte Society for American Archaeology.

Clagett, H. L.
1997. "Household Archaeology at Chan Cahal: A Step Beyond Functionalism," in The Blue Creek Project: Working Papers from the 1996 Field Season. Edited by W. D. Driver, H. L. Clagett, and H. R. Haines, pp. 61-70. San Antonio: Maya Research Program, St. Mary's University.

1999. Paper presented at the Annual Society for American Archaeology Conference. Chicago.

n.d. "Excavations at Chan Cahal," in The Blue Creek Project: Working Papers from the 1997 Season. Edited by W. D. Driver, H. R. Haines, and T. H. Guderjan. San Antonio, TX: Maya Research Program and St. Mary's University.

Clark, J. E., T. A. Lee Jr., and T. Salcedo.
1989. "The Distribution of Obisidian," in Ancient Trade and Tribute: Economies of the Soconusco Area of Mesoamerica. Edited by B. Voohies, pp. 268-284. Salt Lake City: University of Utah Press.

Clark, J. E., and T. A. Lee.
1984. "Formative Obsidian Exchange and the Emergence of Public Economics in Chiapas, Mexico," in Trade and Exchange in Early Mesoamerica. Edited by K. G. Hirth, pp. 235-274. Albuquerque: University of New Mexico Press.

Cliff, M. B.
1982. Lowland Maya Nucleation: A Case Study from Northern Belize. Ph.D., UMI Dissertation Services.

Cobean, R. H.
1991. High-Precision Trace-Element Characterization of Major Mesoamerican Obsidian Sources and Further Analysis of Artifacts from San Lorenzo Tenochtitlan. Latin American Antiquity 2:97-114.

Cobos, R.
1994. "Preliminary Report on the Archaeological Mollusca and Shell Ornaments of Caracol, Belize," in Studies in the Archaeology of Caracol, Belize, vol. 7, Monograph. Edited by D. Z. Chase and A. F. Chase, pp. 139-147. San Francisco: Pre-Columbian Art Research Institute.

Coe, M. D.
1993. The Maya, 5th edition. London: Thames and Hudson.

Coe, M. D., and K. V. Flannery.
1964. The Precolumbian Obsidian Industry of El Chayal, Guatemala. American Antiquity 30:43-49.

Coe, W. R.
1959. "Caches," in Piedras Negras: Artifacts, Caches and Burials, Museum Monographs. Edited by W. R. Coe, pp. 77-79. Philadelphia: University Museum.

1965a. "Caches and Offertory Practices of the Maya Lowlands," in Archaeology of Southern Mesoamerica, vol. 2, Handbook of Middle American Indians. Edited by G. R. Willey, pp. 462-468. Austin: University of Texas Press.

1965b. Tikal -- A Handbook of the Ancient Maya Ruins. Philadelphia: The University Museum, University of Pennsylvania.

1965c. Tikal, Guatemala and Emergent Maya Civilization. Science 147:1401-1423.
1990. Excavation in the Great Plaza, North Terrace, and North Acropolis of Tikal. Vol. 14. Tikal Reports. Philadelphia: University Museum, University of Pennsylvania.

Coggins, C.
1979. "A New Order and the Role of the Calendar: Some Characteristics of the Middle Classic Period at Tikal," in Maya Archaeology and Ethnohistory. Edited by N. Hammond and G. R. Willey, pp. 38-50. Austin: University of Texas Press.

Cole, G. L., and D. A. Andrews-Jones.
1978. Economic Evaluation of the Mineral Potential of Belize: An Overview of hte Anschutz Mining Corporation Exploration Program in Belize, 1975-1977. Anschutz Mining Corporation. Third Annual Report.

Crabtree, D. E.
1968. Mesoamercian Polyhedral Cores and Prismatic Blades. American Antiquity 33:446-478.

Cressey, P. J.
1974. Post-Conquest Developments in the Teotihuacan Valley, Mexico. Part 4: The Early Colonial Obsidian Industry. Vol. 1. Research Report. Iowa: The University of Iowa.

Dreiss, M., and M. Glascock.
1995. "Preliminary Source Analysis of Obisdian from Blue Creek," in Archaeological Research at Blue Creek, Belize. Progress Report of the Third (1994) Field Season. Edited by T. H. Guderjan and W. D. Driver, pp. 99-104. San Antonio: Maya Research Program, St. Mary's University.

Dreiss, M. L.
1986. Obsidian At Colha Belize: A Technological Analysis and Distributional Study Based on Trace Element Data. Masters.

1989. "An Obsidian Distribution Model for the Belize Periphery," in Coastal Maya Trade, vol. 8, Trent University Occasional Papers in Anthropology. Edited by H. McKillop and P. F. Healy, pp. 79-90. Peterborough: Trent University.

Dreiss, M. L., and D. O. Brown.
1989. "Obsidian Exchange Patterns in Belize," in Prehistoric Maya Economies of Belize, Research In Economic Anthropology, Supplement 4. Edited by P. A.

McAnany and B. L. Isaac, pp. 57-90: JAI Press Connecticut.

Dreiss, M. L., D. O. Brown, T. R. Hester, M. D. Glascock, H. Neff, and K. S. Stryker.
1993. Expanding the Role of Trace-Element Studies: Obsidian Use in the Late and Terminal Classic Periods at the Lowland Maya Site of Colha,Belize. Ancient Mesoamerica 4:271-283.

Drennan, R. D.
1984a. Long-Distance Movement of Goods in the Mesoamerican Formative and Classic. American Antiquity 49:27-43.

1984b. Long-Distance Transport Costs in Pre-Hispanic Mesoamerica. American Anthropologist 86:105-112.

Driver, W. D.
1995. "Architecture and Ritual at Structure 1," in Archaeological Research at Blue Creek, Belize. Progress Report of the Third (1994) Field Season. Edited by T. H. Guderjan and W. D. Driver, pp. 27-44. San Antonio: Maya Research Program, St. Mary's University.

1999a. "Excavations in Plaza A," in The Blue Creek Project: Working Papers from the 1997 Season. Edited by W. D. Driver, H. R. Haines, and T. H. Guderjan. San Antonio, Texas: Maya Research Program, St. Mary's University.

1999b. "Structure as Narrative: Monitoring Sociopolitical Change Through Architecture at the Maya Site of Blue Creek, Belize," : Manuscript on file with author.

Drucker, M. D.
1978. The Geology of the Bladen Volcanic Series, Southern Maya Mountains, Belize, Central America. Master's.

Duranni, S., M. T. Khan, and C. Renfrew.
1971. Obsidian Source Identification by Fission Track Analysis. Nature 233:242-245.

Earle, T.
1997. How Chiefs Come to Power: The Political Economy in Prehistory. Stanford, California: Stanford University Press.

Eaton, J. D.
1982. "Colha an Overview of Architecture and Settlement," in Archaeology at Colha, Belize: The 1981 Interim Report. Edited by T. R. Hester, H. J. Shafter, and J. D. Eaton, pp. 11-20. San Antonio: Center for Archaeological Reseach, UTSA.

Ericson, J. E., and J. Kimberlin.
1977. Obsidian Sources, Chemical Characterization and Hydration Rates in West Mexico. Archaeometry 19:157-166.

Ferriz, H.
1985. Caltonac, a Prehispanic Obsidian-Mining Center in Eastern Mexico?: a Preliminary Report. Journal of Field Archaeology 12:363-370.

Ford, A., F. Stross, F. Asaro, and H. V. Michel.
1997. Obsidian Procurement and Distribution in the Tikal-Yaxha Intersite Area of hte Central Maya Lowlands. Ancient Mesoamerica 8:101-110.

Fowler, W. R., Jr., and e. al.
1989. Sources of Obsidian From El Mirador, Guatemala: New Evidence on Preclassic Maya Interaction. American Anthropologist 91:158-168.

Freidel, D. A.
1977. A Late Preclassic Monumental Maya Mask at Cerros, Northern Belize. Journal of Field Archaeology 4:488-491.

1978. "Maritime Adaptation and the Rise of Maya Civilization: The Veiw from Cerros Belize," in Prehistoric Coastal Adaptations: The Ecology and Economy of Maritime Middle America. Edited by B. Stark and B. Voorhies, pp. 239-265. New York: Academic Press.

1979. Culture Areas and Interaction Spheres: Contrasting Approaches to the Emergence of Civilization in the Maya Lowlands. American Antiquity 44:36-54.

1981a. "Civilization as a State of Mind: The cultural evolution of the Lowland Maya," in Transition to Statehood in the New World. Edited by G. D. Jones and R. R. Kautz, pp. 188-227: Cambridge University Press.

1981b. "The Political Economics of Residential Dispersion Among the Lowland Maya," in Lowland Maya Settlement Patterns. Edited by W. Ashmore, pp. 3710382. Albuquerque: University of New Mexico Press.

1986a. "Introduction," in Archaeology at Cerros Belize, Central America: VI - An Interim Report. Edited by R. A. Robertson and D. A. Freidel, pp. xiii-xxiii. Dallas: Southern Methodist University Press.

1986b. "The Monumental Architecture," in Archaeology at Cerros Belize, Central America: VI - An Interim Report. Edited by R. A. Robertson and D. A. Freidel, pp. 1-22. Dallas: Southern Methodist University Press.

Freidel, D. A., and V. Scarborough.
1982. "Subsistence, Trade and Development of the Coastal Maya," in Maya Subsistence. Edited by K. V. Flannery, pp. 131-155. New York: Academic Press.

Freidel, D. A., and L. Schele.
1988a. Kingship in the Late Preclassic Maya Lowlands. American Anthropologist 90:547-567.

1988b. "Symbol and Power: A History of the Lowland Maya Cosmogram," in Maya Iconography. Edited by E. P. Benson and G. G. Griffen, pp. 44-93: Princeton University Press.

Fuller, R. E.
1927. The Mode of Origin of the Color of Certain Varicolored Obsidians. Journal of Geology 35:570-573.

Gann, T.
1900. "Mounds in Northern Honduras," in Nineteenth Annual Report, 1987-1898, Bureau of American Ethnology, vol. Part 2, pp. 661-692. Washington D.C.: Smithsonian Institution.

Gann, T., and M. Gann.
1939. Archaeological Investigations in the Corozal District of British Honduras. Vol. 123. Bulletin of the Bureau of American Ethnology. Washington D.C.: Smithsonian Institution.

Garber, J. F.
1981. Material Culture and Patterns of Artifact Consumption and Disposal at the Maya Site of Cerros in Northern Belize. Ph.D.

1983. Patterns of Jade Consumption at Cerros, Northern Belize. American Antiquity 48:800-807.

Garber, J. F. 1989. The Artifacts. Vol. II. Archaeology at Cerros, Belize. Texas: Southern Methodist University Press.

Garber, J. F., and F. K. Reilly III.
1995. "A Late Formative/Early Classic Mask on Structure B1 at Blackman Eddy, Belize," . San Ignacio, Cayo District, Belize: Paper presented at the 1st International Symposium on Maya Archaeology, Hosted by the Department of Archaeology, Belize.

Gates, W.
1978. An Outline Dictionary of Maya Glyphs. New York: Dover Publications, Inc.

Gerhardt, J. C.
1988. Preclassic Maya Architecture at Cuello, Belize. Vol. 464. BAR International Series. Oxford, UK: British Archaeological Reports (B.A.R.).

Gerhardt, J. C., and N. Hammond.
1991. "The Community of Cuello: The Ceremonial Core," in Cuello: An Early Maya Community in Belize. Edited by N. Hammond, pp. 98-117. Cambridge, UK: Cambridge University Press.

Gibson, E. C.
1982. "An Analysis of the Late Classic Maya Settlement Pattern at Kichpanha, Northern Belize," in Archaeology at Colha, Belize: The 1981 Interim Report. Edited by T. R. Hester, H. J. Shafter, and J. D. Eaton, pp. 152-166. San Antonio: Center for Archaeological Reseach, UTSA.

Gifford, J. C.
1976. Ceramics of Baton Ramie. Vol. 18. Peabody Museum Memoirs. Harvard: Peabody Museum of Archaeology and Ethnology.

Gilgan, E.
1996. "Excavations at the Structure 13 Courtyard Complex," in Archaeological Research at Blue Creek, Belize. Progress Report of the Fourth (1995) Field Season. Edited by T. H. Guderjan, W. D. Driver, and H. R. Haines, pp. 69-77. San Antonio: Maya Research Program, St. Mary's University.

Glascock, M. D.
1994. "New World Obsidian: Recent Investigations," in Archaeometry of Precolumbian Sites., Proceedings of a Symposium Organized by the UCLA Institute of Archaeology and the Getty Conservation Institute. Los Angles: UCLA Department of Archaeology.

1998. "An Overview of Neutron Activation Analysis," , vol. 1998: Missouri University Research Reactor.

Glascock, M. D., J. M. Elam, and R. Cobean. 1988. "Differentiation of Obsidian Sources in Mesoamerica." The 26th International Archaeometry Symposium, 1988, pp. 245-251. Archaeometry '88.

Glascock, M. D., H. Neff, K. S. Stryker, and T. N. Johnson.
1994. Sourcing Archaeological Obsidian by Abbreviated NAA Procedure. Journal of Radioanalytical and Nuclear Chemistry 180:29-34.

Graham, I.
1967. Archaeological Exploration in El Peten, Guatemala. Vol. 33. Middle American Research Institute Publications. New Orleans: Middle American Research Institute, Tulan University.

Graham, J. A., T. R. Hester, and R. N. Jack.
1972. "Sources for the Obsidian at the Ruins of Seibal, Peten, Guatemala," in Studies in the Archaeology of Mexico and Guatemala, vol. 16, Contributions of the

University of California Archaeological Research Facility. Edited by J. A. Graham, pp. 111-116. Berkeley: University of California Archaeological Research Facility, University of California.

1978. "Sources of the Obsdian at the Ruins of Seibal, Peten, Guatemala," in Archaeological Studies of Mesoamerican Obsidian, vol. 3, Studies in Mesoamerican Art, Archaeology and Ethnohistory. Edited by T. R. Hester, pp. 1950199: Ballena Press.

Guderjan, T. H.
1988. Maya Maritime Trade at San Juan, Ambergris Cay, Belize. Ph.D.

1993. Ancient Maya Traders of Ambergris Caye. Benque Viejo del Carmen, Belize: Cubola Books.

1995a. "Blue Creek: A Summary of the First Three Years of Excavation," . San Ignacio, Cayo District, Belize: Paper presented at the 1st International Symposium of Maya Archaeology, Hosted by the Department of Archaeology, Belmopan.

1995b. Maya Settlement and Trade on Ambergris Caye, Belize. Ancient Mesoamerica 6:147-159.

1996a. "The Blue Creek Jade Cache: Early Classic Ritual in Northwestern Belize," : draft paper for "Sowing and Dawning in the Archaeological and Ethnographic Record of Mesoamerica" ed. Shirley Mock.

1996b. "Kakabish: Mapping of a Maya Center," in Archaeological Research at Blue Creek, Belize. Progress Report of the Fourth (1995) Field Season. Edited by T. H. Guderjan, W. D. Driver, and H. R. Haines, pp. 117-120. San Antonio: Maya Research Program, St. Mary's University.

1998. "The Blue Creek Jade Cache: Early Classic Ritual in Northwestern Belize," in The Sowing and the Dawing: Termination, Dedication and Transformation in the Archaeological and Ethnographic Record of Mesoamerica. Edited by S. B. Mock, pp. 101-112. Albuquerque: University of New Mexico Press.

Guderjan, T. H., J. F. Garber, and H. A. Smith.
1989. "Maritime Trade on Ambergris Caye, Belize," in Coastal Maya Trade, vol. 8, Trent University Occasional Papers in Anthropology. Edited by H. McKillop and P. F. Healy, pp. 123-134. Peterborough: Trent University.

Guderjan, T. H., J. F. Garber, H. A. Smith, F. Stross, H. V. Michel, and F. Asaro.
1989. Maya Maritime Trade and Sources of Obsidian at San Juan, Ambergris Caye, Belize. Journal of Field Archaeology 16:363-369.

Guderjan, T. H., H. Haines, M. Lindeman, S. Mock, E. Ruble, F. Salam, and L. Worcester.
1993. Excavations at the Blue Creek Ruin Northwestern Belize: 1992 Interim Report. San Antonio: Maya Research Program.

Guderjan, T. H., and P. J. Weiss.
1995. An Early Classic Ritual at the Blue Creek Ruin. Foundation for the Advancement of Mesoamerican Studies.

Gutchen, M.
1983. The Destruction of Archaeological Resources in Belize, Central America. Journal of Field Archaeology 10:217-226.

Haines, H. R.
1995. "Summary of Excavations at the Temple of the Masks," in Archaeological Research at Blue Creek, Belize. Progress Report of the Third (1994) Field Season. Edited by T. H. Guderjan and W. D. Driver, pp. 73-98. San Antonio: Maya Research Program, St. Mary's University.

1996. "A Preliminary Discussion of Preclassic Midden Deposits at Structure 9," in Archaeological Research at Blue Creek, Belize. Progress Report of the Fourth (1995) Field Season. Edited by T. H. Guderjan, W. D. Driver, and H. R. Haines, pp. 85-108. San Antonio: Maya Research Program, St. Mary's University.

1997. "Continuing Excavations of Preclassic Deposits at Structure 9," in The Blue Creek Project: Working Papers from the 1996 Field Season. Edited by W. D. Driver, H. L. Clagett, and H. R. Haines, pp. 19-24. San Antonio: Maya Research Program, St. Mary's University.

1999. "Continuing Excavations at Structure 9, Blue Creek," in The Blue Creek Project: Working Papers from the 1997 Season. Edited by W. D. Driver and H. R. Haines. San Antonio: Maya Research Program, St. Mary's University.

Haines, H. R., and S. G. Suther.
1997. "A Preclassic Occupation at Structure IA-1," in The Blue Creek Project: Working Papers from the 1996 Field Season. Edited by W. D. Driver, H. L. Clagett, and H. R. Haines, pp. 25-28. San Antonio: Maya Research Program, St. Mary's University.

Haines, H. R., and D. Wilhelmy.
1999. "New Excavations at the Structure 25 Courtyard," in The Blue Creek Project: Working Papers from the 1997 Season. Edited by W. D. Driver and H. R. Haines. San Antonio: Maya Research Program, St. Mary's University.

Hamblin, N. L.
1984. Animal Use by the Cozumel Maya. Tucson, Az: The Univerity of Arizona Press.
Hammond, N. 1972. Obsidian Trade Routes. Science 178:1092-1093.

Hammond, N. (editor)
1973. British Museum - Cambridge University Corozal Project: 1973 Interim Report. Cambridge, UK: Centre of Latin American Studies.

1991b. Cuello: An Early Maya Community in Belize: Cambridge University Press.

Hammond, N.
1976. "Maya Obsidian Trade in Southern Belize," in Maya Lithic Studies: Papers from the 1976 Belize Field Symposium, vol. 4, Special Report. Edited by T. R. Hester and N. Hammond, pp. 71-81. San Antonio: Center for Archaeological Research, University of Texas San Antonio.

1977. The Earliest Maya. Scientific America 236:116-133.

1978. "Cacao and Cobaneros: An Overland Trade Route Between the Maya Highlands and Lowlands," in Mesoamercian Communication Routes and Culture Contacts, Papers of the New World Archaeological Foundation. Edited by T. A. Lee and C. Navarrete, pp. 19-25. Provo: New World Archaeological Foundation.
1983. Nohmul, Belize: 1982 Investigations. Journal of Field Archaeology 10:245-254.

1985. Nohmul: A Prehistoric Maya Community in Belize. Excavations 1973-1983. Vol. 250(i). British Archaeological Reports, International Series. Oxford: BAR.

1989. Obsidian Hydration Dating of the Tecep Phase Occupation at Nohmul. American Antiquity 54:513-521.

1991a. "Archaeological Investigations at Cuello, 1975-1987," in Cuello: An Early Maya Community in Belize. Edited by N. Hammond, pp. 8-22. Cambridge, UK: Cambridge University Press.

1991c. "Obsidian Trade," in Cuello: An Early Maya Community in Belize. Edited by N. Hammond, pp. 197-198. Cambridge: Cambridge University Press.

Hammond, N., A. Clarke, and S. Donaghy.
1995. The Long Goodbye: Middle Precalssic Maya Archaeology at Cuello, Belize. Latin American Antiquity 6:120-128.

Hammond, N., A. Clarke, and C. Robin.
1991. Middle Preclassic Buildings and Burials at Cuello, Belize: 1990 Investigations. Latin American Antiquity 2:352-363.

Hammond, N., M. Nievens, and G. Harbottle.
1984. Trace Element Analysis of Obsidian Artifacts from a Classic Maya Residential Group at Nohmul, Belize. American Antiquity 49:815-820.

Hammond, N., D. Pring, R. Wilk, S. Donaghey, F. P. Saul, E. S. Winger, A. V. Miller, and L. H. Feldman.
1979. The Earliest Lowland Maya? Definition of the Swasey Phase. American Antiquity 44:92-110.

Hansen, R. D.
1990. Excavations in the Tigre Complex, El Mirador, Peten, Guatemala. El Mirador Series, Part 3. Vol. 62. Papers of the New World Archaeological Foundation. Provo, Utah: New World Archaeological Foundation, Brigham Young University.

1992. The Archaeology of Ideology: A Study of Maya Preclassic Architectural Sculpture at Nakbe, Peten, Guatemala. Ph.D., UMI Dissertation Services.

Harrison, P. D.
1999. The Lords of Tikal: Rulers of an Ancient Maya City. New Aspects of Antiquity. London: Thames and Hudson.

Haviland, W. A.
1981. "Dower Houses and Minor Ceremonial Centres at Tikal Guatemala: an investigation into the identification of," in Lowland Maya Settlement Patterns. Edited by W. Ashmore, pp. 89-117. Albuquerque: University of New Mexico Press.

Hay, C.
1977. Use-Scratch Morphology: A Functionally Significant Aspect of Edge Damage on Obsidian Tools. Journal of Field Archaeology 4:491-494.

Hayden, B.
1995. "Pathways to Power: Principles for Creating Socioeconomic Inequalities," in Foundations of Social Inequality, Fundamental Issues in Archaeology. Edited by T. D. Price and G. M. Feinman, pp. 15-86. New York: Plenum Press.

Hayden, B., and A. Cannon.
1983. Where the Garbage Goes: Refuse Disposal in the Maya Highlands. Journal of Anthropological Archaeology 2:117-163.

Healan, D. M.
1993. Local versus non-local obsidian exchange at Tula and its implications for post-Formative Mesoamerica. World Archaeology 24:449-466.

Healy, P. F., H. J. McKillop, and B. Walsh.
1984. Analysis of Obsidian from Moho Cay, Belize: New Evidence on Classic Maya Trade Routes. Science 225:414-417.

Heizer, R. F., H. Williams, and J. A. Graham.
1965. "Notes on Mesoamerican Obsidians and Their Significance in Archaeological Studies," , vol. 1, Contributions of the University of California Archaeological Research Facility, pp. 94-103. Berkeley: University of California Archaeological Research Facility, University of California.

Hester, T. R.
1982. "Archaeology at Colha, Belize: The 1981 Season," in Archaeology at Colha, Belize: The 1981 Interim Report. Edited by T. R. Hester, H. J. Shafter, and J. D. Eaton, pp. 1-10. San Antonio: Center for Archaeological Reseach, UTSA.

Hester, T. R., R. F. Heizer, and R. H. Jack.
1971. "Technology and Geologic Sources of Obsidian Artifacts from Cerro de Las Mesas, Veracruz, Mexico, with observations on Olmec Trade," , vol. 13, Contributions of the University of California Archaeological Research Facility, pp. 133-144. Berkeley: University of Californial Archaeological Research Facility, University of California.

Hester, T. R., R. H. Jack, and R. F. Heizer.
1971. "The Obsidians of Tres Zapotes, Veracruz, Mexico, with observations on Olmec Trade," , vol. 13, Contributions of the University of California Archaeological Research Facility, pp. 65-131. Berkeley: University of Californial Archaeological Research Facility, University of California.

Hester, T. R., and H. Michel.
1980. "Geological Sources of Obsidian Artifacts from the Site of Colha, Belize," in The Colha Project Second Season 1980 Interim Report. Edited by T. R. Hester, J. D. Eaton, and H. J. Shafer, pp. 313-315. San Antonio: Center for Archaeological Research, University of Texas and Centro Studi e Ricerche Ligabue, Venice.

Hester, T. R., and H. J. Shafer.
1984. Exploitation of Chert Resources by the Ancient Maya of Northern Belize, Central America. World Archaeology 16:157-172.

Hester, T. R., and H. J. Shafer. (editors).
1991. Maya Stone Tools. Vol. 1. Monographs in World Archaeology: Prehistory Press.

Hester, T. R., H. J. Shafer, J. D. Eaton, R. E. W. Adams, and G. Ligabue.
1983. Colha's Stone Tool Industry. Archaeology 36.

Holmes, W. H.
1900. The Obsidian Mines of Hidalgo, Mexico. American Anthropologist, n.s. 2:405-416.

Howell, W. K.
1989. Excavations in the Dante Complex at El Mirador, Peten, Guatemala. Vol. 60. Papers of the New World Archaeological Foundation. Provo: Brigham Young University.

Huntley, D. J., and D.C.Bailey.
1978. Obsidian Source Identification by Thermoluminescence. Archaeometry 20:159-170.

Hurcombe, L. M.
1992. Use-Wear Analysis and Obsidian: Theory, Experiments and Results. Vol. 4. Sheffield Archaeological Monographs. Sheffield, England: University of Sheffield.

Jack, R. N., and R. F. Heizer.
1968. "Finger-Printing of Some Mesoamerican Obsidian Artifacts," in Papers on Mesoamerican Archaeology, vol. 5, Contributions of the University of California Archaeological Research Facility, pp. 81-100. Berkeley: University of California Archaeological Research Facility, University of California.

Jack, R. N., T. R. Hester, and R. F. Heizer.
1972. "Geologic Sources of Archaeological Obsidian from Sites in Northern and Central Veracruz, Mexico," in Studies in the Archaeology of Mexico and GuatemalaContributions of the University of California Archaeological Research Facility, vol. 16, Contributions of the University of California Archaeological Research Facility. Edited by J. A. Graham, pp. 117-122. Berkeley: University of California Archaeological Research Facility.

Jackson, T. L., and M. W. Love.
1991. Blade Running: Middle Preclassic Obsidian and the Introduction of Prismatic Blades at La Blanca, Guatemala. Ancient Mesoamerica 2:47-59.

Johnson, J. K.
1976. "Long Distance Obsidian Trade: New Data From the Western Maya Periphery," in Maya Lithic Studies: Papers from the 1976 Belize Field Symposium, vol. 4.

Edited by T. R. Hester and N. Hammond, pp. 83-90. San Antonio: Center for Archaeological Research, UTSA.

Johnson, J. K.
1992. "Obsidian: A Technological Analysis," in Cuello: An Early Maya Community in Belize. Edited by N. Hammond, pp. 169-173: Cambridge University Press.

Kidder, A. V., J. D. Jennings, and E. M. Shook.
1978. Excavations at Kaminaljuyu, Guatemala: Pennsylvania State University.

Kosakowsky, L. J.
1987. Preclassic Maya Pottery at Cuello, Belize. Vol. 47. Anthropological Papers of the University of Arizona. Tucson: The University of Arizona Press.

Lange, F. W.
1971. Marine Resources: A Viable Subsistence Alternative for the Prehistoric Lowland Maya. American Anthropologist 73:619-639.

Leach, B. F., S. Warren, and B. Frankhauser.
1978. Obsidian from the Far North of New Zealand: A method of sourcing based on natural radioactive emission. New Zealand Journal of Science 21:123-128.

Lee, D., and J. J. Awe.
1995. "Middle Preclassic Architecture, Burials, and Craft Specialization: Report on the 1994 Investigations at the Cas Pek Group, Cahal Pech, Belize.," in Belize Valley Preclassic Maya Project: Report on the 1994 Field Season, vol. 10, Occasional Papers. Edited by P. F. Healy and J. J. Awe, pp. 95-115. Peterborough: Trent University.

LeeDecker, C. H.
1994. Discard Behavior on Domestic Historic Sites: Evaluation of Contexts for the Interpretation of House Consumption Patterns. Journal of Archaeological Method and Theory 1:345-375.

Lewenstein, S.
1981. Mesoamerican Obsidian Blades: an Experimental Approach to Function. Journal of Field Archaeology 8:175-188.

1987. Stone Tool Use at Cerros. Austin: University of Texas Press.

Lewenstein, S., and J. Walker.
1984. The Obsidian Chip/Manioc Grating Hypothesis and the Mesoamerican Preclassic. Journal of New World Archaeology 6:25-38.

Lichtenstein, R. J.
1997. "The Structure 19 Courtyard," in The Blue Creek Project: Working Papers from the 1996 Field Season. Edited by W. D. Driver, H. L. Clagett, and H. R. Haines, pp. 39-50. San Antonio: Maya Research Program, St. Mary's University.

Longworth, G., and S. E. Warren.
1979. The Application of Mossbauer Spectroscopy to the Characterisation of Western Mediterranean Obsidian. Journal of Archaeological Science 6:179-193.
Loten, S., and D. M. Pendergast.
1984. Lexicon of Maya Architecture. Vol. 8. Royal Ontario Museum Monographs: Royal Ontario Museum.

Lucero, L. J.
1999. Classic Lowland Maya Political Organization: A Review. Journal of World Prehistory 13:211-263.

MacKinnon, J. J., and S. M. Kepecs.
1989. Prehispanic Saltmaking in Belize: New Evidence. A,merican Antiquity 54:522-533.

Marcus, J.
1983. Lowland Maya Archaeology at the Crossroads. American Antiquity 48:454-488.

Matheny, R. T., R. D. Hansen, and D. L. Gurr.
1980. "Preliminary Field Report, El Mirador, 1979 Season," in El Mirador, Peten, Guatemala: An Interim Report, vol. 45, Papers of the New World Archaeological Foundation. Edited by R. T. Matheny, pp. 1-30. Provo: Brigham Young Unversity.

Maudslay, A. P.
1897. Biologia Centrali-Americana: Archaeology, 1974 Facsimile edition, Milpatron Publishing Corp. edition. Vol. I, II. London: R. H. Porter and Dulau and Co.

Mauss, M.
1990. The Gift: Forms and Reason for Exchange in Archaic Socities. New York: W.W. Norton Press.

Maxwell, D. B. S.
1994. "Marine Materials from Caches and Problematic Deposits at Tikal," . Anaheim, California: Paper Presented at the 59th Annual Meeting of the Society of American Archaeology.

1996. An Analysis of Caches from Four Sites in the Maya Lowlands. Ph.D., UMI Dissertation Services.

1998. "Human Remains as Offerings at Tikal and Uaxactun," : Ms. on file with author.

McAnany, P. A.
1995. Living with the Ancestors: Kinship and Kingship in Ancient Maya Society. Austin: University of Texas Press.

McKillop, H.
1984. Prehistoric Maya Reliance on Marine Resources: Analysis of a Midden from Moho Cay, Belize. Journal of Field Archaeology 11:24-35.

1989. "Coastal Maya Trade: Obsidian Densities at Wild Cane Cay," in Supplement 4, Research in Economic Anthropology. Edited by P. A. McAnany and B. Isaac, pp. 17-56. Greenwich, Connecticut: JAI Press Inc.
1995. "The Role of Northern Ambergris Caye in Maya Obsidian Trade: Evidence from Visual Sourcing and Blade Technology," in Maya Maritime Trade, Settlement, and Populations on Ambergris Caye, Belize. Edited by T. H. Guderjan and J. F. Garber, pp. 163-174. San Antonio, Texas: Maya Research Program and Labyrinthos Press.

1996. Ancient Maya Trading Ports and the Integration of Long-Distance and Regional Economies. Ancient Mesoamerica 7:49-62.

McKillop, H., L. J. Jackson, H. Michel, F. Stross, and F. Asaro.
1988. "Chemical Source Analysis of Maya Obsidian: New Perspectives from Wild Cane Cay, Belize," in Proceedings of the 26th International Archaeometry Symposium. Edited by R. M. Farquhar, R. G. V. Hancock, and L. A. Pavlish, pp. 239-244. Toronto: Archaeometry Laboratory, Department of Physics, University of Toronto.

Michels, J. W.
1975. El Chayal, Guatemala: A chronological and behavioural reassessment. American Antiqutiy 40:103-106.

Mitchum, B. A.
1989. "Obsidian as a Non-Essential Resource," in La Obsidiana en Mesoamerica, Coleccion Cientifica. Edited by M. G. G. and J. E. Clark, pp. 375-378. Mexico D.F.: Instituto Nacional de Antropologia e Historia.

1991. "Lithic Artifacts from Cerros, Belize: Production, Consumption, and Trade," in Maya Stone Tools: Selected Papers from the Second Maya Lithics Conference, vol. 1, Monographs in World Archaeology. Edited by T. R. Hester and H. J. Shafer, pp. 45-53: Prehistory Press.

1994. "Lithic Artifacts from Cerros, Belize: Trade Patterns Among the Late Preclassic Maya," . Atlanta, Georgia: Paper presented at the 93rd Annual Meeting of the American Anthropological Association.

Mock, S. B.
1997. Monkey Business at Northern River Lagoon: A coastal-inland interaction sphere in northern Belize. Ancient Mesoamerica 8:165-183.

1998. "The Defaced and the Forgotten : Decapitation and Flaying/Mutilation as a Termination Event at Colha, Belize," in The Sowing and the Dawing: Termination, Dedication and Transformation in the Archaeological and Ethnographic Record of Mesoamerica. Edited by S. B. Mock, pp. 113-124. Albuquerque: University of New Mexico Press.

Moholy-Nagy, H.
1975. Obsidian at Tikal, Guatemala. Actas de XLI Congresso Internacional de Americanistas 1:511-518.

1976. "Spatial Distribution of Flint and Obsidian Artifacts at Tikal, Guatemala," in Maya Lithic Studies: Papers from the 1976 Belize Field Symposium, Center for Archaeological Research Special Report No 4. Edited by T. R. Hester and N. Hammond, pp. 91-107: Center for Archaeological Research, University of Texas at San Antonio.

1984. Tikal Obsidian: Soures and Typology. American Antiquity 49:104-117.

1989. "Who Used Obsidian at Tikal?," in La Obsidiana en Mesoamerica, Coleccion Cientifica. Edited by M. G. G. and J. E. Clark, pp. 379-389. Mexico D.F.: Instituto Nacional de Antropologia e Historia.
1990. The Misidentification of Mesoamerican Lithic Workshops. Latin American Antiquity 1:268-279.

1994. Tikal Material Culture: Artifacts and Social Structure at a Classic Maya City. Ph.D., UMI Microfilms Dissertation Service International.
Moholy-Nagy, H. 1999. Mexican Obsidian at Tikal, Guatemala. Latin American Antiquity 10:300-313.

Moholy-Nagy, H., and F. W. Nelson.
1987. More Obsidian Source Analysis of Obsidian From Tikal, Guatemala. Lithic Technology 16:13-21.

1990. New Data on Sources of Obsidian Artifacts from Tikal, Guatemala. Ancient Mesoamerica 1:71-80.

Morley, S. G.
1946. The Ancient Maya, 2nd edition. Palo Alto: Stanford University Press.

Neivens, M.
1991. "Operations at the Site of Blue Creek," in Maya Settlement in Northwestern Belize: The 1988 and 1990 Seasons of the Rio Bravo Archaeological Project. Edited

by T. H. Guderjan, pp. 51-54. San Antonio, Texas: Maya Research Program and Labyrinthos Press.

Neivens, M., and D. Libby.
1976. " An Obsidian Workshop at El Posito, Northern Belize," in Maya Lithic Studies: Papers from the 1976 Belize Field Symposium, vol. 4, Center for Archaeological Research Facility, Special Report. Edited by T. R. Hester and N. Hammond, pp. 137-149. San Antonio: University of Texas at San Antonio.

Nelson, F. W.
1985. Summary of the Results of Analysis of Obsidian Artifacts from the Maya Lowlands. Scanning Electron Microscopy II:631-649.
Nelson, F. W., Jr., and D. S. Howard.
1986. Trace Element Analysis of Obsidian Artifacts from El Mirador, Guatemala. Vol. 3. Notes of the New World Archaeological Foundation. Provo, Utah: New World Archaeological Foundation.

Nelson, F. W., K. K. Nielson, N. F. Mangleson, M. W. Hill, and R. T. Matheny.
1977. Preliminary Studies of the Trace Element Composition of Obsidian Artifacts from Northern Campeche, Mexico. American Antiquity 42:209-225.

Nelson, F. W., D. A. Philips, and A. B. Rubio.
1983. "Trace Element Analysis of Obsidian Artifacts from the Northern Maya Lowlands," in Investigations at Edzna, Campeche, Mexico, Vol. 1: Part 1: The Hydraulic Systems, vol. 46, Papers of the New World Archaeological Foundation. Edited by R. T. Matheny, D. L. Gurr, D. W. Forsyth, and F. R. Hauk, pp. 204-219. Provo: New World Archaeological Foundation, Brigham Young University.

Nelson, F. W., R. Sidrys, and R. D. Holmes.
1978. "Trace Element Analysis by X-Ray Fluorescence of Obsidian Artifacts from Guatemala and Belize," in Excavations at Seibal, Department of Peten, Guatemala, Artifacts, vol. 14, Memoirs of the Peabody Museum of Archaeology and Ethnology. Edited by G. Willey, pp. 153-161. Cambridge, MA.

Parsons, L. A., and B. J. Price.
1971. "Mesoamerican Trade and its role in the Emergence of Civilization," in Observations on the Emergence of Civilization in Mesoamerica, vol. 11, Contributions of the University of California Archaeological Research Facility. Edited by R. F. Heizer and J. A. Graham, pp. 169-195. Berkeley: California Archaeological Research Foundation, University of California.

Pastrana, D.
1996. "Excavations of Structures 5 and 6," in Archaeological Research at Blue Creek, Belize. Progress Report of the Fourth (1995) Field Season. Edited by T. H. Guderjan, W. D. Driver, and H. R. Haines, pp. 43-58. San Antonio: Maya Research Program, St. Mary's University.

1999. "An Analysis of the Jades from the Structure 4 Shaft Cache," in The Blue Creek Project: Working Papers from the 1997 Field Season. Edited by H. R. H. W. David Driver. San Antonio, Texas: Maya Research Program.

Pendergast, D. M.
1981. Lamanai, Belize: Summary of Excavation Results, 1974-1980. Journal of Field Archaeology 8:29-53.

1982. Excavations at Altun Ha, Belize, 1964-1970. Vol. 2. Toronto: Royal Ontario Museum.
1986. "Stability Through Change: Lamanai, Belize, from the Ninth to the Seventeenth Century.," in Late Lowland Maya Civilization: Classic to Post Classic. Edited by J. A. Sabloff and E. W. Andrews V., pp. 223-249.

1990. Excavations at Altun Ha, Belize, 1964-1970. Vol. 3. Toronto: Royal Ontario Museum.
PIXE Analytical Laboratories. "The PIXE Technique - It's History and Theory," , vol. 2000: PIXE Analytical Laboratories.

Polanyi, K.
1957. The Great Transformation: The political and economic origins of our time. Boston: Beacon Hill.

Pollock, H. E. D.
1965. "Architecture of the Maya Lowlands," in Archaeology of Southern Mesoamerica, vol. 2, Handbook of Middle American Indians. Edited by G. R. Willey, pp. 378-440. Austin: University of Texas Press.

Pring, D., and M. Walton.
1976. "Survey and Excavation at Cuello," in Archaeology in Northern Belize. British Museum-Cambridge University Corozal Project 1974-1975 Interim Report. Edited by N. Hammond, pp. 138-150. Cambridge, UK: University of Cambridge.
Proskouriakoff, T. 1993. Maya History. Austin: University of Texas Press.

Puleston, D. E.
1983. Tikal Report No. 13: The Settlement Survey of Tikal, 1986 Facsimile Reissue of Original Report edition. Vol. 48. Museum Monograph. Philadelphia: The University Museum.

Rappaport, R. A.
1968. Pigs for the Ancestors: Ritual in the Ecology of a New Guinea People. New Haven, Conn: Yale University Press.

Rathje, W. L.
1972. "Praise the Gods and Pass the Metates: a hypothesis of the development of lowland rainforest civilizations in Mesoamerica," in Contemporary Archaeology, a guide to theory and contributions. Edited by M. P. Leone. Carbondale: Southern Illinois University.

Rathje, W. L., D. A. Gregory, and F. M. Wiseman.
1978. "Trade Models and Archaeological Problems: Classic Maya Examples," in Mesoamerican Communication Routes and Cultural Contacts, vol. 40, Papers of the New World Archaeological Foundation. Edited by T. A. Less and C. Naverrete, pp. 147-175. Provo, Utah: New World Archaeological Foundation.

Rathje, W. L., and J. A. Sabloff.
1973. Ancient Maya Commercial Systems: A Research Design for the Island of Cozumel, Mexico. World Archaeology 5:221-231.

Reeves, R. D., and G. C. Armitage.
1973. Density Measurements and Chemical Analysis in the Identification of New Zealand Archaeological Obsidians. New Zealand Journal of Science 16:561-572.

Renaud, J.
1999. "A Summary of Archaeological Investigations at Structure 24," in The Blue Creek Project: Working Papers from the 1997 Season. Edited by W. D. Driver, H. R. Haines, and T. H. Guderjan, pp. 47-48. San Antonio, Texas: Maya Research Program.

Rice, D. S.
1976. Middle Preclassic Maya Settlement in the Central Maya Lowlands. Journal of Field Archeology 3:425-445.

Rice, P.
1984. Obsidian Procurement in the Central Peten Lakes Region, Guatemala. Journal of Field Archaeology 11:181-194.

Rice, P., H. V. Michel, F. Asaro, and F. Stross.
1985. Provenience Analysis of Obsidian from the Central Peten Lakes Region, Guatemala. American Antiquity 50:591-604.

Robertson, R. A., and D. A. Freidel. (editors).
1986. An Interim Report. Vol. I. Archaeology at Cerros Belize, Central America. Dallas, Tx: Southern Methodist University Press.

Romney, D. H., A. C. S. Wright, R. H. Arbuckle, and V. E. Vial.
1959. Land in British Honduras, Report of the Land Use Survey Team. London: Her Majesty's Stationary Office.

Rovner, I.
1989. "Patrones Anomalos en la Importacion de Obsidiana en el Centro de las Tierras Bajas Mayas," in La Obsidiana en Mesoamerica, Coleccion Cientifica. Edited by M. G. G. and J. E. Clark, pp. 369-373. Mexico D.F.: Instituto Nacional de Antropologia e Historia.

Sabloff, J. A., and W. L. Rathje.
1975. A Study of Changing Precolumbian Commerical Systems. The 1972-1973 Seasons at Cozumel, Mexico. Vol. 3. Monographs of the Peabody Museum. Cambridge: Peabody Museum of Archaeology and Ethnology, Harvard University.

Sanders, W. T.
1973. "The cultural ecology of the Lowland Maya: a reevaluation," in The Classic Maya Collapse. Edited by T. P. Culbert, pp. 325-365. Albuquerque: University of New Mexico Press.

1977. "Environmental Heterogeneity and the Evolution of Lowland Maya Civilization," in The Origins of Maya Civilization. Edited by R. E. W. Adams, pp. 287-297. Albuquerque: University of New Mexico Press.

Santley, R. S., J. M. Kerley, and R. R. Kneebone.
1986. "Obsidian Working, Long-Distance Exchange, and the Politico-Economic Organization of Early States in Central Mexico," in Economic Aspects of Pre-Hispanic Highland Mexico, Research in Economic Anthropology, Supplement 2. Edited by B. L. Isaac, pp. 101-132. Greenwich, Conneticut: JAI Press.

Schele, L., and D. Freidel.
1990. A Forest of Kings: The Untold Story of the Ancient Maya. New York: Quill William Morrow.

Schele, L., and P. Mathews.
1991. "Royal Visits and other intersite relationships among the Classic Maya," in Classic Maya Political History: Hieroglyphic and Archaeological Evidence. Edited by T. P. Culbert, pp. 226-252. Cambridge: Cambridge University Press.

Schiffer, M. B.
1972. Archaeological Context and Systemic Context. American Antiquity 37:156-165.

1976. Behavioural Archaeology. New York: Academic Press.

1977. "Towards a Unified Science of the Cultural Past," in Research Strategies in Historical Archaeology. Edited by S. South, pp. 13-40. New York: Academic Press.

Sharer, R.
1994. The Ancient Maya, 5 edition: Stanford University Press.

Sheets, P., K. Hirth, F. Lange, F. Stross, F. Asaro, and H. V. Michel.
1990. Obsidian Sources and Elemental Anlayses of Artifacts in Sourthern Mesoamerica and the Northern Intermediate Area. American Antiquity 55:144-158.

Sheets, P. D.
1975. A Reassessment of the Precolumbian Obsidian Industry of El Chayal, Guatemala. American Antiquity 40:98-103.

Sheets, P. D., and G. R. Muto.
1972. Pressure Blades and Total Cutting Edge: An experiment in Lithic Technology. Science 175:632-634.

Sidrys, R.V.
1979. Supply and Demand Among the Classic Maya. Current Anthropology 20:594-597.

1983. "Exchange Systems," in Archaeological Excavations in Northern Belize, vol. 17, Monograph. Edited by R. V. Sidrys, pp. 375-389. Los Angeles: UCLA Institute of Archaeology, University of California Press.

Sidrys, R., J. Andresen, and D. Marcucci.
1976. Obsidian Sources in the Maya Area. Journal of New World Archaeology 1:1-13.

Sidrys, R. V., and J. Kimberlin.
1979. Use of Mayan Obsidian Through Time: Trace-Element Data from El Balsamo, Guatemala. Jouranl of Field Archaeology 6:116-122.

Smith, A. L.
1950. Uaxactun, Guatemala: Excavations of 1931-1937. Vol. 588. Carnegie Institute of Washington Publications. Washington D. C.: Carnegie Institute of Washington.
1972. Excavations at Altar de Sacrificios: Architecture, Settlement, Burials and Caches. Vol. 64 (2). Papers of the Peabody Museum of Archaeology and Ethnology. Cambridge: Harvard University.

Smith, H. A., and K. M. McField.
1996. Excavations at Lamanai South, Orange Walk District, Belize: Second Interim Report of Field Activities. Department of Archaeology.

Spence, M. W.
1982. "The Social Context of Production and Exchange," in Contexts for Prehistoric Exchange, Studies in Archaeology. Edited by J. E. Ericson and T. K. Earle, pp. 173-197: Academic Press.

1996. Commodity of Gift: Teotihuacan Obsidian in the Maya Region. Latin American Antiquity 7:21-39.

Spence, M. W., and J. Parsons.
1967. Prehispanic Obsidian Mines in Southern Hidalgo. American Antiquity 32:542-543.
Stanchly, N.
in press. "An Analysis of the Faunal Remains Recovered from Blue Creek, Belize, 1996 and 1997 Field Seasons," in The Blue Creek Project: Working Papers from the 1997 Field Season. Edited by W. D. Driver and H. R. Haines. San Antonio: Maya Research Program, St. Mary's Texas.

Stevenson, D. P., F. H. Stross, and R. F. Heizer.
1971. An Evaluation of X-Ray Fluorescence Analaysis as a Method for Correlating Obsidian Artifacts with Source Locations. Archaeometry 13:17-25.

Stross, F. H., H. V. Michel, F. Asaro, and R. Gruhn.
1977. Sources of Some Obsidian Flakes from a Paleoindian Site in Guatemala. American Antiquity 42:114-118.

Stross, F. H., D. P. Stevenson, J. R. Weaver, and G. Wyld.
1971. "Analysis of Ancient American Obsidians by X-Ray Fluorescence and Neutron Activation Analysis," in Science and Archaeology. Edited by R. H. Brill, pp. 210-221. Cambridge: MIT Press.

Stross, F. H., J. R. Weaver, G. E. A. Wyld, R. F. Heizer, and J. A. Graham.
1968. Analysis of American Obsidians by X-Ray Fluorescence and Neutron Activation Analysis. Contributions of the University of California Archaeological Research Facility 5:59-79.

Sullivan, L. A.
1991. Preclassic Domestic Architecture at Colha, Belize. Masters, University of Texas at Austin.
Thompson, J. E. S. 1965. "Archaeological Synthesis of the Southern Maya Lowlands: An Introduction," in Archaeology of Southern Mesoamerica, vol. 2, Handbook of Middle American Indians. Edited by G. R. Willey, pp. 331-359. Austin: University of Texas Press.

Thompson, J. E. S.
1966. The Rise and Fall of Maya Civilization, second edition. Norman: University of Oklahoma Press.

1990. Maya History and Religion. Norman: University of Oklahoma Press.

Tourtellot, G.
1978. "Getting What Comes Unnaturally: on the Energetics of Maya Trade," in Papers on the Economy an Architecture of hte Ancient Maya, vol. VIII, Monograph. Edited by R. Sidrys. Los Angeles: Institute of Archaeology, University of California.

1988. Excavations at Seibal, Department of Peten, Guatemala. Peripheral Survey and Excavation. Vol. 16. Memoirs of the Peabody Museum of Archaeology and Ethnology. Cambridge: Harvard University.

Trik, A. S.
1963. The Splendid Tomb of Temple 1 at Tikal, Guatemala. Expedition 6:2-18.

Tylor, E. B.
1861. Anahuac: or Mexico and the Mexicans, Ancient and Modern. London: Longman, Green, Longman and Roberts.

Vail, G.
1988. The Archaeology of Coastal Belize. Vol. S463. BAR International Scries. Oxford, UK: British Archaeological Reports.

Valdez Jr., F., and S. B. Mock.
1991. Additional Considerations for Prehispanic Saltmaking in Belize. American Antiquity 6:520-525.

Walker, C.
1999. "PIXE - Proton Induced X-ray Emission," , vol. 2000: UK ESCA.

Walker, W. H.
1995. "Ceremonial Trash?," in Expanding Archaeology. Edited by J. M. Skibo, W. H. Walker, and A. E. Nielsen, pp. 67-79. Salt Lake City: University of Utah Press.

Weaver, J. R., and F. H. Stross.
1965. "Analysis by X-Ray Fluorescence of Some American Obsidians," , vol. 1, Contribution of the University of California Archaeological Research Facility, pp. 89-93. Berkeley: University of California Archaeological Research Facility, University of California.

Webster.
1950. *Webster Dictionary of English*. New Jersey: Castle Book Press.

Weiner, A. B.
1985. Inalienable Wealth. American Ethnologist 12:210-227.

1992. Inalienable Possessions: the Paradox of Keeping While Giving: University of California Press.

Weiss, P. J.
1995. "Architecture and Ritual at Structure 4," in Archaeological Research at Blue Creek, Belize. Progress Report of the Third (1994) Field Season. Edited by T. H. Guderjan and W. D. Driver, pp. 45-62. San Antonio: Maya Research Program, St. Mary's University.

1996. "The Continuation of Excavations at Structure 4," in Archaeological Research at Blue Creek, Belize. Progress Report of the Fourth (1995) Field Season. Edited by T. H. Guderjan, W. D. Driver, and H. R. Haines, pp. 37-42. San Antonio: Maya Research Program, St. Mary's University.

Whitehouse, R. D. (editor).
1983. The Facts on File Dictionary of Archaeology. New York: Facts on File Publications.

Willey, G. R.
1972. The Artifacts of Altar de Sacrificios. Vol. 64(1). Papers of the Peabody Museum of Archaeology and Ethnology: Peabody Museum of Archaeology and Ethnology.

Williams-Beck, L. A.
1997. Blue Creek Complete Vessels. San Antonio: Unpublished Manuscript on File with Maya Research Program.

Wright, A. C. S., D. H. Romney, R. H. Arbuckle, and V. E. Vial.
1959. Land in British Honduras, Report of the Land Use Survey Team. London: Her Majesty's Stationary Office.

Yaeger, J. R.
1992. Settlement Survey at Gallon Jug, Orange Walk District, Belize. Masters.

Zeitlin, R. N.
1982. Toward a More Comprehensive Model of Inter-Regional Commodity Distribution: Political Variables and Prehistoric Obsidian Procurement in Mesoamerica. American Anitquity 47:260-275.

www.ingramcontent.com/pod-product-compliance
Lightning Source LLC
LaVergne TN
LVHW071214210926
848482LV00015B/810
*9781407309095*